Introduction to
ANIMAL PHYSIOLOGY

Introduction to
ANIMAL
PHYSIOLOGY

Ian Kay

Department of Biological Sciences, Manchester Metropolitan University, Manchester, UK

βIOS
SCIENTIFIC
PUBLISHERS

© BIOS Scientific Publishers Limited, 1998

First published 1998

Reprinted 2002 by Pear Tree Press Ltd, Stevenage Herts, SG1 2BH

Transferred to Digital Printing 2006

A CIP catalogue record for this book is available from the British Library.

ISBN 1 85996 046 4

BIOS Scientific Publishers Ltd
9 Newtec Place, Magdalen Road, Oxford OX4 1RE, UK
Tel. +44 (0)1865 726286. Fax +44 (0)1865 246823
World Wide Web home page: http://www.bios.co.uk/

Published in the United States of America, its dependent territories and Canada by Springer-Verlag New York Inc., 175 Fifth Avenue, New York, NY 10010-7858, in association with BIOS Scientific Publishers Ltd.

Published in Hong Kong, Taiwan, Singapore, Thailand, Cambodia, Korea, The Philippines, Indonesia, The People's Republic of China, Brunei, Laos, Malaysia, Macau and Vietnam by Springer-Verlag Singapore Pte. Ltd, 1 Tannery Road, Singapore 347719, in association with BIOS Scientific Publishers Ltd.

Production Editor: Fran Kingston.
Typeset and illustrated by The Florence Group, Stoodleigh, UK.

Contents

Preface   viii
Abbreviations   x

1 **Homeostasis: The central concept in physiology**   **1**
An introduction to animal physiology   1
The body fluids of animals   2
Homeostasis   3
Homeostatic control systems – feedback   4
Feedforward   7
Non-physiological homeostatic mechanisms   7
Acclimatization   8

2 **Neurons and nervous systems**   **9**
Introduction   9
Cellular elements of the nervous system   9
How neurons work   13
Synaptic transmission   20
Organization of nervous systems   26

3 **Receptors and effectors**   **33**
Introduction   33
Sensory neurophysiology   33
Classification of sensory receptors   34
Sensory receptor function   34
Sensory reception   37
Effectors – the responses to sensory information   43
Intracellular movement   45
Amoeboid movement   46
Muscle and movement   47
Skeletal systems   52

4 **Endocrine function**   **55**
Introduction   55
Definition of endocrine systems   55
Identification of endocrine organs   57

The chemical nature of hormones 58
The mechanism of hormone action 59
Invertebrate endocrine systems 62
Vertebrate endocrine systems 67

5 **Ventilation and gas exchange** 73
Introduction 73
Gases in air and water 74
A comparison of air and water as respiratory media 75
Gas exchange by simple diffusion across the general body surface area 76
The evolution and design considerations of gas exchange organs 77
Gills 79
Lungs 81
Tracheal systems 85
Control of ventilation 87

6 **Thermoregulation in animals** 91
The importance of temperature to animal physiology 91
Classification of temperature regulation 92
Heat exchange interactions between animals and the environment 92
Ectotherms 95
Endotherms 99
Control of body temperature in endotherms 104
A comparison of ectothermy with endothermy 105

7 **Circulatory systems** 107
Functions of circulatory systems 107
The composition of blood 108
The heart 111
Types of circulatory system 114
Transport of oxygen 119
Transport of carbon dioxide 123

8 **Gastrointestinal function** 125
Introduction 125
Feeding mechanisms 125
The need for a gastrointestinal system 131
Generalized structure and function of gastrointestinal systems 132
Excretion and water absorption 145

9 **Osmoregulation** 147
Introduction 147
The principles of osmosis 147
Generalized osmotic responses of animals 149
The osmotic responses of animals 151

10 **Excretory mechanisms** 161
The need for excretory organs 161
Types of excretory organs 161
Nitrogen excretion 171

11 **Reproduction** **175**
  Introduction 175
  Asexual reproduction 175
  Sexual reproduction 177
  Gamete production 178
  Fertilization 183
  Development and pregnancy 185
  Birth 189
  Lactation 190
  Maternal behavior 191

  Appendix. An outline classification of animals **193**
  Glossary **197**
  Index **211**

Preface

This book has been developed from a series of lectures given to first year students at Manchester Metropolitan University, reading for a degree in Biological Sciences. It is intended to provide an easy introduction to the subject of animal physiology at first year university level. At this level, students want to build on and develop the knowledge base they have gained prior to entering university. However, at this stage they do not wish to be inundated and confused with detailed knowledge about the minutia of physiological processes. It is hoped that this book will fall between these two areas and serve as a suitable introductory text.

Many students consider physiology to be the study of mammalian and, in particular, human physiology. This book has deliberately taken a comparative view of physiology. At this stage in their biological careers, students simply require a thorough understanding of the key principles of physiology, and these principles are best learned using examples drawn from a wide range of animals. By understanding these principles, a solid foundation is formed for further physiological studies. The book also serves to reinforce another very important principle: the study of 'simple' animals can provide useful insights into how higher animals (including humans) work. The earliest work on the squid giant axon and the first investigations into the way that neurons work are good examples of this.

Writing this book has been an enjoyable learning experience in itself. Any errors or omissions remain my fault alone and I would be pleased to learn of any. Equally, if you have enjoyed the book I would like to hear from you.

I am grateful to colleagues within the Department of Biological Sciences at Manchester Metropolitan University and former students, who made helpful comments on draft versions of the text. Thanks must also go to the staff at BIOS for their help and advice – in particular, Rachel Offord and Fran Kingston, who have endured so many 'it'll be with you soon' letters, faxes, e-mails and telephone messages! There are several other people to whom acknowledgement must be made – without them the book would not have been possible. They are my parents, my wife Jan and children Matthew and Amy. I can now get back to being a normal son, husband and dad, free from the shackles of writing.

Ian Kay

Abbreviations

ACTH	adrenocorticotropic hormone
ADH	antidiuretic hormone
ADP	adenosine diphosphate
ANP	atrial naturetic peptide
ATP	adenosine triphosphate
BAT	brown adipose tissue
2,3-BPG	2,3-bisphosphoglycerate
cAMP	cyclic adenosine monophosphate
DAG	diacyl glycerol
ECF	extracellular fluid
EPSP	excitatory postsynaptic potential
FSH	follicle-stimulating hormone
GABA	γ-aminobutyric acid
GDP	guanosine diphosphate
GnRH	gonadotropin-releasing hormone
GTP	guanosine triphosphate
5-HT	5-hydroxytryptamine (serotonin)
ICF	intracellular fluid
IP_3	inositol trisphosphate
IPSP	inhibitory postsynaptic potential
LH	luteinizing hormone
MIH	molt-inhibiting hormone
p	partial pressure
P_i	inorganic phosphate
PIP_2	phosphatidyl inositol bisphosphate
PTTH	prothoracicotropic hormone
RMP	resting membrane potential
STP	standard temperature and pressure
TMAO	trimethylamine oxide
TRH	thyrotropin-releasing hormone
TSH	thyrotropin-stimulating hormone

Homeostasis: The central concept in physiology

1.1 An introduction to animal physiology

Physiology is 'the science of normal functions and phenomena of living things'; it concerns itself with how animals work. This science of normal function can be investigated at several different levels, from the cellular level through the functioning of discrete organs to the functioning of the whole animal. Comparative animal physiology looks at the way different animals deal with common problems, such as how oxygen can be delivered to cells to meet their metabolic needs and how animals cope with changes in their immediate external environment; for example, how do estuarine invertebrates deal with the constant changes in salinity to which they are exposed?

The external environment probably presents the biggest challenge that an animal is likely to face. External environments can be divided into two categories: terrestrial and aquatic. It should be remembered that some animals may live in both environments, perhaps at different stages in their development, for example, amphibians which are fully aquatic during their larval stage, but which are semi-terrestrial when adult. Generally, however, an animal's external environment is rarely constant. For example, there will be changes in temperature, water availability, gas concentrations and so on. These changes, which may occur on a daily or seasonal basis, represent a challenge to the normal functioning of the animal. The fact that the external environment is changing means that the body fluids, e.g. **blood plasma**, **hemolymph** and **extracellular fluid**, which constitute the internal environment and which surround the cells of the animal are constantly being challenged. An absolute requirement of animals is that their internal conditions remain constant, or as near constant as possible. This maintenance of a constant internal state is known as **homeostasis**. The fluid that surrounds animal cells in many cases has a composition which is quite different from that of the external environment that surrounds the animal. Obviously, in the case of terrestrial animals, the fluid (which may be either liquid or gaseous) which surrounds them is air. In this case, comparisons are made between their body fluids and sea water, since this is thought to be where terrestrial animals emerged from during evolution. What animals must strive to do is to maintain their body fluids in a relatively constant state in terms of, for example, ionic composition, pH, dissolved gas concentrations, nutrient levels and so on. Obviously, then, homeostasis is a major theme running through animal physiology. Before looking at the types of mechanisms and processes involved in the maintenance of homeostasis, it is necessary to consider the composition of the fluids which are contained within animals.

1.2 **The body fluids of animals**

As indicated above, the body fluids of animals must be held relatively constant, in terms of ionic composition, pH, etc., if they are to avoid being at the mercy of their environments. The ability of animals to maintain internal fluids of a relatively constant composition has allowed them to colonize new environments during their evolutionary history, e.g. the transition from aquatic environments to the terrestrial environment. The transition from aquatic to terrestrial environments has only occurred in two groups of animals – the insects and the vertebrates – to any great extent, and in other groups, such as spiders and some molluscs, to a lesser extent. Clearly, the ability to colonize new environments like this opens up all sorts of new possibilities in the evolution of animals. Body fluids are divided into two categories, each of which will now be discussed.

1.2.1 *Extracellular fluids*

The **extracellular fluid (ECF)** is the fluid that surrounds the cells of any particular animal. In the case of the protozoans (considered here as single-cell animals) the ECF is the fluid of their external environment. The ECF of some simple multicellular marine animals is more-or-less identical to the environment in which they live – for example, some jellyfish have a solute composition identical to seawater with the exception that their sulfate concentration is only half that of seawater. The reason for this discrepancy is that the sulfate ion is dense (relative to other ions). This means that alterations to the concentration of sulfate ions in body fluids will have significant effects on the overall density of the animal and, therefore, its buoyancy. Life is thought to have originated in the oceans, so it is not unreasonable that such simple animals should have a similar composition to seawater. However, more advanced marine animals, freshwater and terrestrial animals all maintain an ECF that is quite different in composition from that of seawater. The precise composition of the ECF will vary from animal to animal, but some generalizations may be made. The main extracellular cation is almost always Na^+ and the main extracellular anion is Cl^-. A cation is a positively charged ion, which moves towards the negative pole (cathode) in an electric field. An anion is a negatively charged ion that will move towards the positive pole (anode) in an electric field. The regulation of body fluid composition in animals is considered in more detail in Chapter 10.

1.2.2 Intracellular fluids

As the name implies, **intracellular fluid (ICF)** is the fluid contained within animal cells. By definition, it must have the same total concentration of solutes as the ECF has. Without this the cell would not be in osmotic equilibrium with its surroundings and would be at risk of gaining or losing water and the dangers associated with this. For example, if a cell gains water it will swell, leading to the rupture of the cell membrane and cell death. However, rather like the ECF and external environment, the ECF and ICF need not have identical compositions (indeed this is the case and much metabolic energy is expended in maintaining this difference) as long as solute concentrations are equal. The composition of the ICF will, again, vary from animal to animal, but the major cation in the cell is K^+. In contrast to the ECF, overall electrolyte concentrations are lower and there are large quantities of proteins that are generally associated with negative charges. The pH of the ICF is slightly alkaline. Such conditions are found almost universally in all animals.

1.3 **Homeostasis**

As described previously, animals must ensure that the ICF and ECF remain as stable as possible – gas concentrations must be held at the correct level, the pH and concentrations of solutes must be constant and so on. This maintenance of a constant internal environment is called homeostasis.

The concept of homeostasis has its origins in nineteenth-century French physiology. It was the French physiologist Claude Bernard who realized the importance of stability in the internal environment. His classic statement of 1857 sums up well this absolute requirement of animals:

> 'It is the fixity of the *milieu interieur* which is the condition of free and independent life, and all the vital mechanisms, however varied they may be, have only one object, that of preserving constant the conditions of life in the internal environment.'

The *milieu interieur* to which Bernard refers is the internal environment of the animal. However, it was several decades later that the term homeostasis was first used. It was coined by the American physiologist Walter Canon in 1929. He stated

> 'This "internal environment" as Claude Bernard called it, has developed as organisms have developed; and with it there have evolved

remarkable physiologic devices which operate to keep it constant. . . . So long as this personal, individual sack of salty water, in which each one of us lives and moves and has his being, is protected from change, we are freed from serious peril. I have suggested that the stable state of the fluid matrix be given the name homeostasis'.

Homeostasis is the central theme in physiology. There are a huge number of examples of homeostasis, including the maintenance of the correct levels of electrolytes and dissolved gases in body fluids, the maintenance of the optimum pH of body fluids, and so on. Indeed, it is extremely difficult to think of many aspects of animal physiology which do not impinge upon this concept of homeostasis. As animals have become more complicated and specialized during the course of evolution, so homeostasis has become more important to their physiology. Physiology could be viewed, rather teleologically, as the means by which homeostasis is maintained. Accordingly, much of our knowledge of homeostatic mechanisms comes from the higher animals, particularly mammals, and in many of the lower invertebrates homeostasis is less apparent (i.e. fewer aspects of the animal's physiology are controlled and regulated). Many such animals do not maintain an internal environment that is different from their external. Such animals are said to **conform**, and any changes in the external environment are mirrored in the internal environment. However, there are limits to the degree of change that can be tolerated; excessive changes will result in severe disruption and possibly death. This is also the case in animals that maintain an internal environment that differs from the external environment. The internal environment in this case is said to be **regulated**: homeostatic mechanisms maintain this difference and its constancy.

1.4 Homeostatic control systems – feedback

Perhaps the most common homeostatic control system is based around the principle of feedback. Feedback is of two sorts: negative and positive. **Negative feedback** can be defined as the change of a variable that is opposed by responses which tend to reverse this change. For example, in birds and mammals which maintain a constant body temperature, a rise in body temperature produces responses which return body temperature to its original, desired value. Thus, negative feedback leads to stability in physiological systems. This contrasts with **positive feedback** systems where an initial change in a particular variable results in further change. By and large, the role of positive feedback in the maintenance of homeostasis is minimal. One possible example is the process of blood coagulation. This process operates via a positive feedback mechanism and it could be considered to be involved in the maintenance of a constant circulating blood volume. In many other cases, the involvement of positive feedback mechanisms in controlling, or attempting to control the normal physiological functions of an animal would be disastrous. In the example of temperature control cited above, consider what would happen if this system was under the influence of positive feedback. Suppose body temperature had increased, if this initial disturbance was then subjected to positive feedback the result would be a further increase in body temperature. However, there are other biological examples of positive feedback, for example, the functioning of nerve cells. In this case, the initial influx of Na^+ ions during the early stages of an **action potential** produces a depolarizing response that results in further Na^+ entry. This causes further depolarization and influx of more Na^+ ions. This process will be discussed in more detail in Chapter 2. In general, though, examples of biological processes that are subject to positive feedback are very few.

The components of a feedback system are shown in *Figure 1.1*. There are three principal components: a **receptor**, an integrating center and an **effector** or effectors. The receptor is responsible for detecting change in the environment of the animal, either the external environment in which it lives, or its internal environment (e.g. a change in environmental temperature or intracellular pH). As will be seen later, there are a large number of **receptors**, each monitoring a particular aspect of the environment. The function of the receptor is to convert the detected change in the environment into action potentials which are sent via the afferent division of the nervous system to the integrating center. The integrating center is usually the brain or spinal cord in those animals that possess one. The role of the integrating center is to 'compare' the incoming information about a particular variable with what the variable should be. For example, the **hypothalamic** region of the brain is the integrating center for the control of body temperature in mammals. On the basis of the incoming information, in this case from thermoreceptors, the hypothalamus 'decides' what appropriate responses must be initiated to restore body temperature to its desired value. This response is brought about by the action of effectors, which

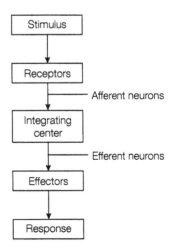

Figure 1.1. The basic components of a feedback system and their arrangement.

are stimulated via efferent (motor) neural pathways. Effector is the general term given to structures which bring about a biological response. Such responses may include muscular, neural or **endocrine** activation. An example of the functioning of a homeostatic control system is shown in *Figure 1.2*.

Clearly, from the above description, the integrating center, whichever organ or tissue it may be, must have a predetermined value for each variable that it controls. This value is known as the set point and it is the value of a particular variable that the animal strives to maintain. In the case of body temperature, for mammals the set point is approximately 37°C. For other animals this value will vary; some birds attempt to maintain a body temperature as high as 42°C, whilst other animals are unable to maintain a constant body temperature. The subject of thermoregulation is considered in Chapter 6. It can be seen from *Figure 1.2* that homeostatic control is really a balance between input and output. In the case of the thermoregulatory system, it is a balance between heat gain and heat loss. The set point is a hypothetical concept; more realistically, there is a range within which the variable is acceptable. In the case of body temperature in humans, the set point is 37°C; however, actual body temperature is acceptable if it lies within ±1°C of this value. Every other physiological variable will have its own range over which it may vary. For example, plasma (the fluid component of blood) pH is held between 7.35 and 7.45 and plasma K^+ concentration is held between 3 and 5.5 mmol l^{-1}. Since nothing is held at a constant level (albeit the fluctuations around the set point of a particular variable are small), homeostasis is never really achieved in the literal sense of its meaning. The actual range which is permitted will vary for different variables. This introduces the hierarchical nature of homeostasis, by which it is meant that some variables are more

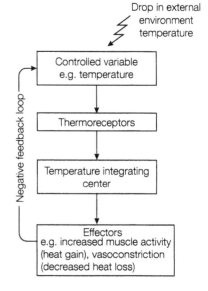

Figure 1.2. The operation of a negative feedback loop regulating body temperature. An initial drop in body temperature is detected by thermoreceptors, which initiate a response that returns temperature to the desired level.

closely controlled than others. This implies that those variables that are most tightly controlled are more important. An example of a variable that is controlled very closely is pH of blood and other body fluids. The reason for this is that even small changes in pH would have severe effects on enzyme (i.e. protein) structure and function. Remember that enzymes are composed of a chain of amino acids whose final structure is maintained by a variety of 'weak' chemical bonds. One such type of 'weak' bond found in enzymes is the ionic interaction between individual amino acids. Changes in pH would alter the ionization status of the chemical groups involved in these ionic interactions, which would lead to disruption of the bonds and, therefore, disruption of the final structure of the enzyme. Obviously, this would have a disastrous effect on the animal's metabolic processes. In contrast, O_2 levels in mammalian blood are less closely controlled – O_2 levels may decrease by 30–40% before any ventilatory effects are noticed (assuming that all other variables remain constant). The fact that control has a hierarchical nature also implies that the various homeostatic control systems work cooperatively, rather than in isolation. For example, consider an animal living in the desert. During the day, there may be times when it is heat-stressed (i.e. in danger of its body temperature rising too high). One way to overcome this would be to lose water by evaporation which would have a cooling effect. However, this now produces a conflict – the need to reduce body temperature versus the need to conserve water. This is discussed in Chapter 9.

1.5 Feedforward

Although negative feedback is an important mechanism for maintaining homeostasis, there are other physiological methods by which control of the internal environment may be achieved. Perhaps the most important of these is the process of feedforward. Feedforward is anticipatory activity – behavior which serves to minimize disruption to physiological systems before it happens. A good example of this is eating and drinking at the same time. Eating has a potentially dehydrating effect; the increased osmolarity of the contents of the gut may promote the loss of water from body fluids to maintain **osmotic equilibrium**. To minimize this disruption to the homeostasis of body fluids, many animals drink at the same time as they eat. There are also other aspects of animal behavior which will contribute to the maintenance of homeostasis. For example, animals can learn to avoid foodstuffs that make them vomit, which has a disruptive effect on homeostasis.

1.6 Non-physiological homeostatic mechanisms

The homeostatic mechanisms described thus far have a basis in the physiology of the animal, requiring some sort of regulatory mechanism (e.g. thermoregulation, pH regulation). However, it is possible to achieve some sort of control without requiring physiological control mechanisms. This can be demonstrated in many aquatic species, vertebrates as well as invertebrates, living in large bodies of water, the temperature of which changes very little. The body temperature of the animal mirrors that of the water in which it lives, and, assuming that the water temperature changes very little, the animal's body temperature also barely changes. This type of homeostasis is called **equilibrium homeostasis**. In essence, the animals are simply conforming to the temperature of the external environment. Whether this is true homeostasis or not is a debatable issue. This system cannot be employed for all aspects of a particular animal's physiology, with the exception of some of the simplest aquatic invertebrates.

1.7 Acclimatization

So far, homeostasis has been described in terms of physiological variables, each of which has a range within which it must be maintained. Superimposed upon this view of homeostasis is the ability of animals to alter the range over which a particular variable is maintained. This ability to alter the range is called **acclimatization**. It can be considered as a fine tuning of homeostatic control systems, the cause of which is some factor in the environment. In this way, regulation is the product of the basic control system together with environmental influences on a particular variable (an analogy of the situation would be the phenotype of an organism being a product of its genotype and the environment in which it lives). For example, the physiology of animals which live at sea level differs from that of animals of the same species which live at high altitude because of the different amounts of oxygen available at each level. The availability of oxygen decreases with increasing altitude. Thus, humans living at altitude show a variety of physiological (and anatomical) adaptations when compared with their sea level counterparts, such as changes in the sensitivity of the receptors which monitor blood O_2 levels, structural differences in the blood vessels that carry deoxygenated blood back to the lungs, and differences in the number and function of red blood cells.

Further reading

Hardy, R. (1983) *Homeostasis.* Edward Arnold, London.
Langley, L. L. (1973) *Homeostasis – Origins of the Concept.* Dowden, Hutchinson and Ross, Pennsylvania.

Neurons and nervous systems

2.1 Introduction

The evolution of animals from simple unicellular organisms to complex multicellular organisms allowed the development of specialized, dedicated organs and structures, such as digestive systems and circulatory systems. It released cells from having to be a jack of all trades. However, it also presented these animals with a new problem: it was now necessary to be able to control and coordinate the activities of many different cells. Failure to do so would defeat the advantages of a multicellular way of life. Two systems have evolved to ensure that the activities of any animal are efficiently controlled and coordinated – these are the nervous and endocrine systems. In this chapter, the function and organization of nervous systems will be discussed. The nature of endocrine control will be described in Chapter 4.

2.2 Cellular elements of the nervous system

2.2.1 Neurons

The nervous system is made up of two different cell types: neurons and glial cells. **Neurons** are the workhorses of the nervous system. They work by generating and conducting **action potentials**, which are simply changes in the polarity of the potential difference (voltage) that exists across the membrane of a neuron (see Section 2.3.2). Action potentials represent the means by which information is transmitted throughout the nervous system and, therefore, the means by which it achieves control and coordination. The structure of a typical neuron is shown in *Figure 2.1*.

In reality, neurons take on many differing appearances and some examples of the forms they take are shown in *Figure 2.2*. However, irrespective of the actual appearance of an individual neuron, they all share common characteristics. A typical neuron consists of a dendritic

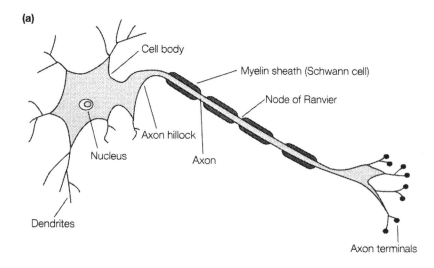

(a)

Cell body

Myelin sheath (Schwann cell)

Node of Ranvier

Axon hillock

Nucleus

Axon

Dendrites

Axon terminals

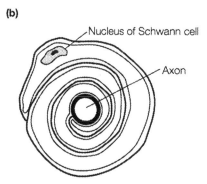

(b)

Nucleus of Schwann cell

Axon

Figure 2.1. (a) The main structural features of a typical myelinated neuron. Action potentials pass from the cell body to the axon terminals. Unmyelinated neurons are 'naked'. (b) The Schwann cell wraps itself around the axon in several concentric circular layers to form the myelin sheath.

region, a somatic region and an axonic region. **Dendrites** are processes extending from the soma or cell body of the neuron. Their function is to receive information from other neurons and from sensory receptors, the latter providing information about what is happening in the animal's environment. In some situations the dendrites may be a sensory structure in their own right – for example, in receptors which are responsible for the detection of touch and pressure in skin. Any information that dendrites receive, which may occur, for example, via the binding of a **neurotransmitter** released from another neuron or through physical changes in their structure (e.g. physical deformation of pressure receptors), is converted into changes in membrane potential (see later) which are transmitted towards the cell body of the neuron. The cell body, sometimes called the soma, contains all the typical organelles of any cell (e.g. a nucleus and mitochondria). This is the region where some neurotransmitters, in particular the

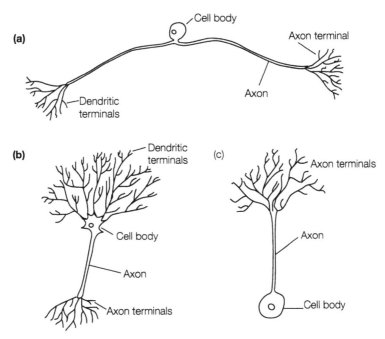

Figure 2.2. The neuron may take a variety of forms. (a) Mammalian sensory neuron; (b) mammalian cerebellar neuron; (c) invertebrate motor neuron.

neuropeptides, which are made of proteins, are synthesized and transported to the axon terminal where they are released during the process of synaptic transmission. Originating from the cell body is another projection called the axon. The junction between the cell body and the **axon** is called the **axon hillock**. This region is of special significance as this is the site of origin of action potentials. The function of the axon is to transmit action potentials down its length to the axon terminal. The axon terminal connects with a variety of other structures, including dendrites, cell bodies, axons of other neurons, or non-neural tissue (e.g. muscle or glandular tissue). There is no physical contact between the axon terminal and other structures since there is a gap between them called the **synapse**. In some cases the axon is enveloped by a **myelin sheath**, which is formed by a group of glial cells called **Schwann cells** which wrap themselves around the axon. The sheath is discontinuous, and the gaps in it are called **nodes of Ranvier**. The function of the myelin sheath is to increase the velocity at which an action potential passes down the length of the axon (see Section 2.3.3). Under normal conditions, the passage of action potentials along an axon occurs in one direction only, from the cell body towards the axon terminals.

It is possible to classify neurons using either structural or functional criteria. On a structural basis, classification is determined by the number of processes originating from the cell body. Adopting this approach, three types of neuron may be seen: multipolar neurons, bipolar neurons and unipolar neurons. The appearance of each of these types of neuron is shown in *Figure 2.3*. Similarly, neurons can be divided into three functional classes. These are sensory, or afferent, neurons, motor, or efferent, neurons and, connecting them together, inter, or internuncial, neurons. The arrangement of these neurons is shown in *Figure 2.4*.

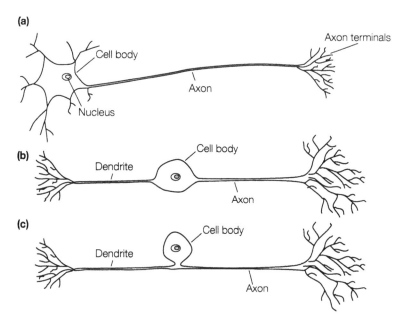

Figure 2.3. The classification of neurons based on the number of processes or extensions originating from the cell body. (a) Multipolar; (b) bipolar; (c) unipolar.

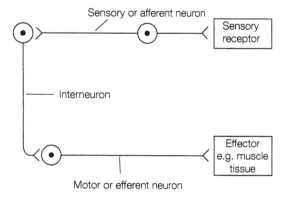

Figure 2.4. The classification of neurons on a functional basis. Neurons may be divided into three functional subtypes: sensory (afferent), motor (efferent) and interneurons.

2.2.2 Glial cells

The second group of cells seen in the nervous system of animals are the **glial cells**. Glial cells are intimately associated with neurons, although they themselves do not conduct action potentials. In the human brain, for example, glial cells outnumber neurons by about 10:1. The function of glial cells is one of supporting the work of neurons. For example, they may provide neurons with nutrients, ensure that the ionic environment surrounding them is correctly maintained, and remove waste materials. There are several types of glial cell, including astrocytes, oligodendrocytes and microglia. Perhaps one of the best known types of glial cells are the Schwann cells that wrap themselves around the axon of neurons to form the myelin sheath.

2.3 How neurons work

As mentioned previously, neurons work by the generation and conduction of action potentials, which are waves of electrical activity that pass along neurons. This occurs because the membrane of a neuron is electrically unstable, which means that the potential difference that exists across the neuronal cell membrane can transiently change. Before the generation and transmission of action potentials are considered, it is necessary to look at the electrical activity of the neuronal cell membrane at rest.

2.3.1 The resting membrane potential

Neurons, like other cells, have a potential difference (voltage) across their membranes. This potential difference is called the **resting membrane potential** (RMP). This voltage can be measured using intracellular electrodes, as shown in *Figure 2.5*. When measured experimentally, the RMP is generally of the order of –75 mV, the inside of the neuron being negative with respect to the outside. It should be remembered that neither the inside nor the outside of the neuron are charged solutions – there is no inherent potential difference either inside or outside the neuron. The potential difference exists between the two regions of the neuron. The membrane potential could be compared with the potential of a battery.

The RMP can be easily understood if we consider the distribution of ions across the neuronal cell membrane together with the differential permeability of the neuronal cell membrane to various ions. *Table 2.1* below lists the major ionic components of intracellular and extracellular fluid (ICF and ECF).

Table 2.1 shows that there is an unequal distribution of ions across the neuronal cell membrane. This is true for other ions and also for other cell types. It is important to note that at rest the neuronal cell membrane

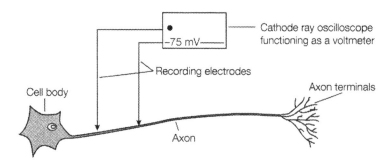

Figure 2.5. The measurement and recording of the resting membrane potential using intracellular electrodes. One of the pair of electrodes penetrates the cell whilst the other remains outside the cell. The voltage measured across the cell membrane is approximately −75 mV (inside is negative with respect to the outside) and is termed the resting membrane potential.

Table 2.1. A table showing the typical concentrations of the principal ions in the ECF and ICF of mammals

	[Ion] mM	
Ion	ECF	ICF
Na^+	155	15
K^+	4	125
Cl^-	100	8

is highly permeable to K^+ ions. This is because, at rest, K^+ channels in the cell membrane are open and permit the passage of K^+ ions. The membrane is far less permeable to Na^+ ions – only about one-twenty-fifth that of K^+. The role played by Na^+ permeability in determining the RMP is discussed later in this section. The permeability to ions is determined by the presence of selective **ion channels** in the cell membrane, 'selective' meaning that only one particular type of ion may move through the channel. Thus, there are K^+ channels which only allow K^+ ions to move through them, Na^+ channels which only allow Na^+ ions to move through them, and so on. A channel can be thought of as a protein pore which spans the cell membrane, and which may either be closed or open. When a particular channel is open, movement of ions across the membrane may occur, and when the channel is closed, movement of ions is prevented.

We first need to consider the role of the K^+ ion in determining the RMP. Because of the unequal distribution of K^+ across the cell membrane, there is a large, outwardly-directed K^+ concentration gradient. This means that K^+ will tend to move out of the cell (the cell membrane is fully permeable to this ion due to the presence of open K^+ channels). As K^+ ions move out of the neuron, they take positive charge with them, and the inside of the neuron becomes negatively charged (remember that both the ICF and ECF are electrically neutral, containing the same total number of positive and negative charges). The number of K$^+$ ions which actually generate this potential difference across the membrane is very small. The loss of K$^+$ ions across the membrane cannot continue indefinitely because the build-up of positive charge on the outside of the neuron (due to K$^+$ ions which have already left the neuron) will prevent further movement of K$^+$ ions outwards as like charges repel. At this point, there will be a dynamic equilibrium between the movement of K$^+$ out of the cell down its concentration (chemical) gradient, and the movement of K$^+$ back into the cell down its electrical gradient (remember the inside of the neuron is now slightly negatively charged). This situation is called an **electrochemical equilibrium** – the electrical and concentration gradients are now equal and opposite to each other. Therefore, at this point there is no further net movement of K$^+$ ions. The potential difference at which this equilibrium exists is called the potassium equilibrium potential. This potential can be calculated using the Nernst equation, which is shown below.

$$E_K = \frac{RT}{zF} \log_e \frac{[K^+]_{out}}{[K^+]_{in}}$$

Where

E_K	=	Potassium equilibrium potential (in volts)
R	=	universal gas constant
z	=	charge on the ion
F	=	Faraday's constant
T	=	absolute temperature (°K)
$\log_e$	=	natural log
$[K^+]_{out}$	=	extracellular concentration of potassium
$[K^+]_{in}$	=	intracellular concentration of potassium

At 37°C this equation simplifies to

$$E_K = 0.0267 \log \frac{[K^+]_{out}}{[K^+]_{in}}$$

What the equation is telling us is the potential difference that will develop across a membrane which is freely permeable to potassium, and which separates two unequal concentrations of potassium ions, at the point of electrochemical equilibrium.

If we calculate the potassium equilibrium potential for a neuron, using the concentrations of potassium in *Table 2.1*, we obtain a value of approximately –90 mV. This is quite close to the experimentally recorded value of –75 mV. The reason for the mismatch between the theoretical value for the RMP and the experimentally recorded value is that other ions, in addition to K$^+$, participate in the generation of the RMP. The one we need to consider is the Na$^+$ ion. Previously, it was mentioned that the neuronal membrane was far less permeable to Na$^+$ ions than it was to K$^+$ ions. In reality, the permeability of the neuronal cell membrane to Na^+ ions is only about one-fiftieth to one-hundredth of its permeability to K^+ ions. However, even this small influx of Na^+ into the interior of the neuron, down its concentration gradient, delivers positive charge into the neuron. The result of this is that the RMP is shifted from its theoretical value of –90 mV to the experimentally recorded value of –75 mV, due to the injection of positive charge from Na^+ ions.

The Goldman equation, a derivation of the Nernst equation, takes into account the distribution (i.e. the internal and external concentrations) and also the permeability of several different ions at the same time, rather than individually, as in the Nernst equation. When the membrane equilibrium potential is calculated (it is termed the membrane equilibrium potential because several ions are now being considered) using this equation, the value obtained is essentially identical to the value of RMP obtained experimentally. There is one other important aspect to the RMP that has yet to be considered, and that is the Na^+/K^+-ATPase pump. This is a membrane pump which pumps out three Na^+ ions in exchange for the entry of two K^+ ions. The pump is described as being electrogenic since it exchanges unequal amounts of charge between the interior and exterior of the neuron. The fact that it pumps out more positive charge than it pumps in, resulting in a net accumulation of negative charge within the neuron, does make a minor contribution to the generation of the RMP, but it is certainly not the cause of the RMP. However, its role is vital as it ensures that the concentration gradients of K^+ and Na^+ are maintained. Without these gradients, there would be a gradual decline in the RMP since it is the ability of small numbers of K^+ ions to leave the cell which results in the generation of the RMP. A simple analogy would be that of a battery running down and eventually becoming flat.

2.3.2 The action potential

Action potentials are transient changes in membrane polarity, when the interior of the neuron changes from being negatively charged to being positively charged for a few milliseconds. The shape of an action potential when recorded with an intracellular electrode is shown in *Figure 2.6*. Action potentials are the means by which neurons 'convey' information. They are transmitted down the entire length of the axons of neurons, in some cases at velocities of up to 120 ms^{-1}. This velocity of conduction only occurs in large, myelinated axons. In smaller, nonmyelinated axons, the velocity of action potential conduction may only reach 2.5 ms^{-1}. One other important aspect of the action potential is that it is transmitted without decrement along the entire length of the axon, so that the size of the action potential at the axon hillock is exactly the same size as the action potential that appears at the axon terminal. What then is the ionic basis of the action potential?

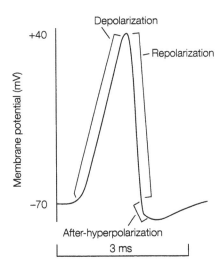

Figure 2.6. The phases of an action potential as recorded with an intracellular electrode.

The action potential reflects the opening of channels, and the influx of Na$^+$ ions into the interior of a neuron following a stimulus of some sort, e.g. electrical stimulation when investigating the action potential experimentally, or activation from another neuron in 'real life'. Not all stimuli will generate an action potential – only stimuli which are able to bring the RMP to a threshold level will suffice. This **action potential threshold** is a membrane potential some 10–15 mV above that of the RMP. At threshold, a positive feedback mechanism is initiated, whereby the initial entry of Na$^+$ ions leads to the further entry of Na$^+$ ions and so on (*Figure 2.7*). This initial change in the membrane potential is known as depolarization. The fact that Na$^+$ is responsible for depolarization can be checked by calculating the Na$^+$ equilibrium potential from the Nernst equation. By using the values for Na$^+$ concentrations in the ECF and ICF from *Table 2.1* in the Nernst equation, a value of +60 mV will be obtained. This coincides with the peak voltage generated during an action potential. Almost as soon as the Na$^+$ channels have opened, they close, and a further series of K$^+$ channels opens. This causes the membrane potential to begin to return to its resting value (the RMP) – this is termed repolarization and results in the

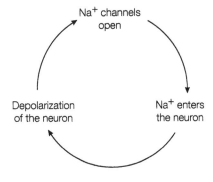

Figure 2.7. The positive feedback cycle that occurs at threshold and ensures rapid depolarization and generation of the action potential.

removal of positive charge, in the form of K⁺ ions, from the interior of the neuron. The K⁺ channels that have opened during the action potential (and which are quite distinct from those that are open at rest) are relatively slow to close, which means that too much positive charge is lost from the neuron. This is termed the after-hyperpolarization and is indicated on the diagram of the action potential shown in *Figure 2.6*. This overshoot is corrected by the Na⁺/K⁺-ATPase pump which exchanges Na⁺ for K⁺ as described previously.

2.3.3 Transmission of the action potential along the axon

Action potentials which originate in the axon hillock have to pass to the axon terminal before they can influence other neurons, muscle or glandular tissue. They do so by the generation of **local currents**. This process is depicted in *Figure 2.8*. What happens is that during an action potential the positive charge on the inside of the axon membrane is drawn towards the adjacent, negatively charged region of membrane at rest. The influx of positive charge tends to move the membrane potential of this region towards threshold. Once threshold is reached, an action potential is generated. This continues all the way down the axon to its terminal. The reason that myelination increases the speed at which action potentials are transmitted is that instead of each immediately adjacent region of membrane having to become depolarized, the action potential jumps from one node of Ranvier to the next. This is called saltatory conduction. The fact that the axon is myelinated means that the movement of ions necessary for the production of action

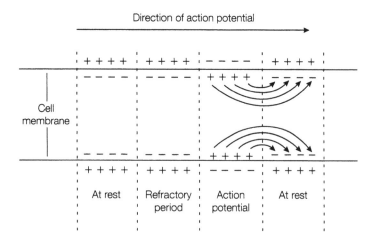

Figure 2.8. Generation of local currents in the transmission of an action potential. The flow of positive charge from a depolarized region to adjacent regions at rest generates an action potential. The preceding region of the membrane is in its refractory period so the action potential can only be transmitted in one direction.

potentials is prevented, except at 'naked' regions of the membrane. Therefore, the only places that the required ion movements may occur are at the nonmyelinated nodes of Ranvier and, thus, the action potential appears to jump from one node to the next down the length of the axon. In this way, the velocity with which the action potential is transmitted down the length of the axon increases. A simple analogy would be walking a certain distance in less time by taking large strides as opposed to small steps.

2.3.4 Signaling in the nervous system

In Section 2.3.2 the ionic basis of the action potential was described. It was stated that an action potential will only be generated when the membrane potential reaches a certain threshold. Once generated, action potentials are of a constant magnitude. For example, if an electrical stimulus of say 1 V will generate an action potential in a neuron, then the action potential generated by a stimulus of 10 V will be exactly the same size. Thus, action potentials generated by neurons are said to be 'all or nothing' – they are either present or they are not. It is impossible to, say, have a double sized action potential or half an action potential. How then does the nervous system of an animal convey information? For example, in humans, how does a mildly painful stimulus differ from an extremely painful stimulus? The answer is in the frequency with which action potentials are generated. A mildly painful stimulus may evoke 10 action potentials per second, whilst the extremely painful stimulus may generate 100 action potentials per second. This means, then, that information in the nervous system is frequency coded. Given that an action potential lasts for 2–3 ms in total, the maximum frequency of action potentials is about 300–500 per second. In part, the maximal rate of generation of action potentials is limited by the refractory period of the neuron. The **refractory period** can be divided into two components. The first is the absolute refractory period. During this time it is impossible to generate a second action potential in the neuron due to the inactivation of the Na^+ channels, which are essential for action potential production. The second component is the relative refractory period. In this case, it is possible to generate a second action potential in the neuron, but to do so requires an increased stimulus (compared to that which generated the first). The explanation for the relative refractory period is that at the end of the action potential, K^+ channels are still open and positive charge is leaving the cell (remember the inside of the neuron must become positively charged during an action potential). However, during this stage the Na^+ channels are able to open and thus depolarize the cell. In essence, then, there has almost been a temporary increase in threshold. Any depolarizing response (i.e. Na^+ entry) must be able to overcome the hyperpolarizing response due to K^+ loss.

2.4 Synaptic transmission

At the end of the last century, there was much debate as to whether individual neurons were discrete structures or whether, by means of physical contact, they were continuous with each other. The discovery and acceptance that they were discrete structures was due to the work of y Cajal and Golgi, who won the Nobel Prize in Medicine and Physiology for this work in 1906. The fact that they were discrete structures meant that individual neurons must communicate with other neurons. How was this achieved? The presence of a synapse, a term introduced by the English physiologist Sherrington to describe the gap between two neurons, now opened up the debate as to the mechanism by which information (i.e. an action potential) could be transmitted between neurons. Two potential mechanisms were suggested: electrical and chemical transmission. Each of these will now be discussed.

2.4.1 Electrical transmission across synapses

Electrical synapses, or **ephapses,** as they are sometimes called, are by far the simplest mechanism by which an action potential can be transferred from one neuron to another. However, they are far less common than chemical synapses, particularly in the higher animals. The structure of a typical electrical synapse is shown in *Figure 2.9*. In this situation, the pre- and postsynaptic membranes lie close to each other, forming a specialized cell–cell contact known as a gap junction. The gap junction is sometimes called a **connexon**, and consists of a protein structure linking the membranes of the two cells. This allows ions for example, to pass from the presynaptic neuron to the postsynaptic neuron. The gap junction may open and close, thus allowing or preventing an action potential to pass from one neuron to the next. It is also possible that some electrical synapses allow action potentials to travel in one direction only, from the presynaptic to the postsynaptic neuron. The synapse relies on the formation of local currents as described previously. Because of this, small, presynaptic neurons are unable to stimulate larger, postsynaptic neurons (including muscle cells). Hence, the use of electrical synapses is not as widespread as the use of chemical synapses. Electrical synapses have been demonstrated in several invertebrate phyla including the annelid worms, arthropods and molluscs. They are also known to occur in vertebrates, being involved, for example, in the escape reaction of fish. When startled, fish flex their bodies, flip their tails and swim away from danger. This response is mediated, in part, by a group of neurons known as Mauthner cells. These cells are interneurons which innervate motor neurons, which, in turn, innervate the muscles of the body wall, the main muscles used for swimming. The electrical synapse is important here because it ensures that there is close synchronization and activation of the muscles, allowing the fish to escape from danger.

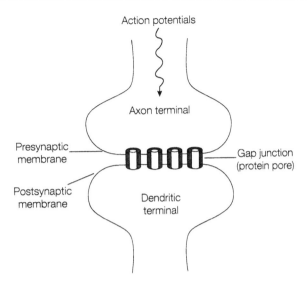

Figure 2.9. Organization of an electrical synapse (ephapse). The pre- and postsynaptic membranes lie close to each other and are physically connected by gap junctions. Ions pass from the presynaptic to the postsynaptic membrane via protein pores (connexons).

2.4.2 Chemical transmission across synapses

Far more common than electrical transmission is chemical transmission at synapses. The structure of a typical synapse is shown in *Figure 2.10*. Depolarization of the presynaptic terminal results in an influx of Ca^{2+} into the presynaptic terminal. The Ca^{2+} enters through ion channels which have opened in response to depolarization. Such Ca^{2+} channels are an example of a voltage-gated channel, i.e. one whose state of opening changes in response to alterations in the membrane potential. The effect of Ca^{2+} influx is to activate, amongst others, the enzyme calcium/calmodulin-dependent protein kinase 1. This enzyme, like any other kinase, phosphorylates substrates. In the case of this particular enzyme, the substrate phosphorylated is synapsin. Ordinarily, synapsin is attached to the vesicle which contains the neurotransmitter substance. When phosphorylated, synapsin detaches from the vesicle allowing the vesicle to fuse with the presynaptic membrane, probably at specific fusion points on the membrane of the presynaptic axon terminal. By a process of **exocytosis**, the neurotransmitter is released into the synapse where it diffuses across and combines with a specific receptor on the postsynaptic membrane. Once the neurotransmitter has combined with its receptor, it is able to influence the membrane potential of the post-synaptic neuron. It may do this in one of two ways:

(i) The receptor may form part of a larger ion channel/receptor complex. Thus, when the receptor is activated by the neurotrans-mitter, this may lead to conformational changes in the structure of the ion channel. This in turn will open the channel, allowing

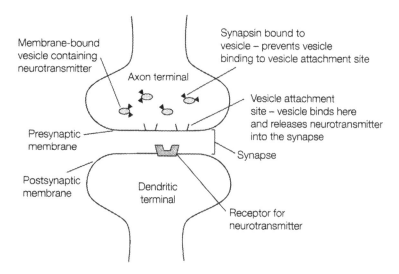

Figure 2.10. The basic elements of the chemical synapse. It is also possible for synapses to occur between axons, between axons and cell bodies and between axons and effector organs (e.g. muscle).

the passage of ions across the membrane and leading to a change in the membrane potential;

(ii) The receptor may, when activated, produce a **second messenger** molecule, e.g. cAMP. This second messenger may in turn influence the state of opening of an ion channel and, therefore, alter the membrane potential.

A list of some of the substances which are known to function as neurotransmitters is shown in *Table 2.2*.

As can be seen, a whole variety of substances are known to act as neurotransmitters. In order to be classed as a neurotransmitter, a substance needs to meet several criteria. Some of the criteria which must be met are outlined below.

(i) The candidate molecule must be synthesized in the neuron from which it is released.

(ii) The candidate molecule must be stored within the neuron from which it is thought to be released.

(iii) Presynaptic stimulation must result in the release of the candidate molecule.

(iv) Application of the candidate molecule at the appropriate postsynaptic site must produce the same postsynaptic response as presynaptic stimulation.

(v) Agents which block the postsynaptic response produced by presynaptic stimulation, should also block the response when the candidate molecule is applied exogenously.

(vi) There must be rapid metabolism and breakdown of the candidate molecule when it is released from axon terminals.

Table 2.2. Some established neurotransmitters found in animal nervous sytems

Substance	Animal
Acetylcholine	Flatworms, insects, molluscs, vertebrates – extremely widespread
Amines	
Dopamine	Molluscs, crustaceans, vertebrates
Noradrenaline (norepinephrine)	Cnidarians, molluscs, vertebrates
Serotonin (5-HT)	Arthropods, annelids, vertebrates
Histamine	Arthropods, vertebrates
Excitatory amino acids	
Glutamate	Crustaceans, insects, vertebrates
Aspartate	Crustaceans, vertebrates
Inhibitory amino acids	
γ-aminobutyric acid (GABA)	Annelids, vertebrates
Peptides [a]	
Substance P	Vertebrates
Vasopressin (ADH)	Vertebrates
Short cardiac peptide	Molluscs
FMRFamide[b]	Molluscs, arthropods, vertebrates
Proctolin	Molluscs, annelids
Purines	
Adenosine	Sipunculans

[a] There are many proteins and peptides which are thought to act as neurotransmitters or neuromodulators. Some of them are specific to a particular group of animals and do not appear in any other animals.
[b] These are a group of tetrapeptide molecules with examples in many phyla. In some instances, the molecule that is present is closely related to the FMRFamide. FMRF relates to the amion acid sequence in the peptide – Phe-Met-Arg-Phe.

Many other substances are thought to be neuroactive compounds if they fail to meet the criteria for neurotransmitters but still influence synaptic function. In the majority of cases, these substances are stored and released from the same presynaptic terminal as the neurotransmitter. Such compounds are called neuromodulators. Their function is to modulate the activity of the neurotransmitter. For example, they may alter the binding of the neurotransmitter to its receptor on the postsynaptic cell, change the number of receptors located on the postsynaptic membrane, or influence the amount of neurotransmitter released from the presynaptic terminal. Quite often, the distinction between neurotransmitter and neuromodulator is blurred.

Having stimulated a postsynaptic neuron, a neurotransmitter must be rapidly inactivated. This must be done to prevent excessive stimulation of the postsynaptic cell. For example, consider a presynaptic neuron innervating a muscle cell. Continuous activation by the released neurotransmitter would result in the muscle cell going into a state of maintained contraction. There are several ways in which neurotransmitters are removed from a synapse, but probably the simplest way is metabolism of the substance in the synapse. For example, the

neurotransmitter acetylcholine is metabolized by the enzyme acetyl-
cholinesterase as follows:

acetylcholine → acetate + choline

The acetate re-enters the circulation and choline is actively transported
back into the presynaptic neuron where it can be resynthesized into
acetylcholine. Another way of removing neurotransmitters from the
synapse is to transport them into cells and break them down intracel-
lularly. A good example of this is the neurotransmitter noradrenaline.
This can be taken up into either neurons or non-neural tissue (e.g. glial
cells) and metabolized to inactive products via a series of enzyme-
catalyzed reactions.

2.4.3 Activation of the postsynaptic cell – postsynaptic potentials

The type of channel opened as a result of the neurotransmitter combining
with its receptor will determine whether the postsynaptic cell is excited
or inhibited. For example, if the channel opened is selective for Na^+ ions,
these ions will flow into the cell, bringing with them positive charge and
producing a depolarizing (i.e. excitatory) response. If, on the other hand,
the channel opened is selective for K^+ ions, then these will leave the
neuron, leaving behind a net negative charge and thus producing a
hyperpolarizing (i.e. inhibitory) response. It should be stressed that
the effect of opening channels does not result in the production of
action potentials *per se*. The changes in membrane potential generated
here are called **postsynaptic potentials**. Excitatory postsynaptic potentials (EPSP) are those which depolarize and inhibitory postsynaptic
potentials (IPSP) are those which hyperpolarize. These potentials must
be transmitted down dendrites and across the cell body to the axon
hillock. Only if they bring the axon hillock region to **threshold** will an
action potential be generated. Obviously, hyperpolarizing responses
will inhibit action potential production. The difference between action
potentials and local potentials is that **local potentials** decrease in size
from their point of origin. This is explained in *Figure 2.11*.

In reality, any one postsynaptic neuron may have hundreds of presynaptic neurons connecting to it. What the postsynaptic cell must do
is integrate all the information it receives, and, on the basis of this,
an action potential or action potentials will be generated. Of these
synapses, some will be excitatory whilst others will be inhibitory, some
will be near the axon hillock (and therefore have more chance of influencing membrane potential there) whilst others will be much further
away from the axon hillock on distant dendrites (with less chance of
influencing the axon hillock membrane potential). This is further
complicated by the frequency with which individual inputs to a particular neuron are activated. One important aspect of the production of
postsynaptic potentials is that they may be summated, unlike action
potentials which are all or nothing, as described earlier. The two types

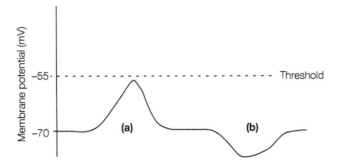

Figure 2.11. The time course of local potentials. (a) An excitatory post-synaptic potential (EPSP). Although a depolarizing response, the threshold is not reached and no action potential is generated. (b) An inhibitory post-synaptic potential (IPSP). This is a hyperpolarizing response.

of **summation** are illustrated in *Figure 2.12*. Temporal summation is the summation of various sub-threshold stimuli received in quick succession from the same input. The threshold is reached and an action potential is generated. Spatial summation occurs when various stimuli from different inputs are received simultaneously and combine to reach the threshold at which an action potential is generated. In both types of summation, a single stimulus would not be sufficient to generate an

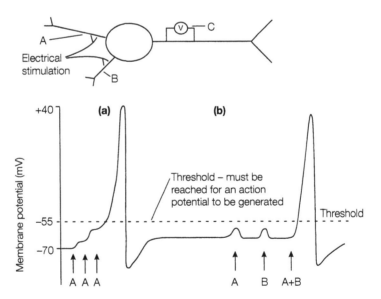

Figure 2.12. Two types of summation. A single neuron is electrically stimulated through one or both dendrites at A and B. The response is recorded at C. The trace recorded with intracellular electrodes is shown below. (a) Temporal summation: three subthreshold stimuli are delivered in rapid succession via input A. Threshold is reached and an action potential is generated. (b) Spatial summation: stimuli from inputs A and B are delivered simultaneously. Threshold is reached generating an action potential.

action potential. Summation can occur with inhibitory stimuli as well as excitatory stimuli. Summation of inhibitory stimuli causes the membrane potential to become increasingly more negative.

2.5 Organization of nervous systems

Nervous systems can be simply described as aggregations or collections of neurons which are arranged to work in a coordinated function. At the simplest level, a nervous system need only be formed from one neuron which has a sensory dendritic function and whose axon terminals synapse with some sort of effector cell (e.g. muscle cells). This provides the animal with the capability to respond to changes in either its internal or its external environment. What follows is a brief description of the arrangement of nervous systems in the animal kingdom.

2.5.1 Unicellular animals

It is perhaps confusing to talk of unicellular animals having a nervous system because, by definition, they are the nervous system, just as they are the circulatory, respiratory and excretory systems. However, they still require the ability to control and coordinate their activities. Consider for example a ciliated protozoan, such as *Paramecium*. The cilia that cover its body surface are used for several things; for example, they are used for locomotion of the animal and also used in feeding, where they waft trapped food particles back to the mouth (*Figure 2.13*).

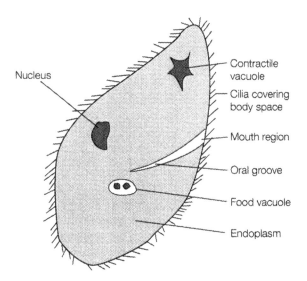

Figure 2.13. Diagram of a protozoan. The entire body surface is covered with cilia which can be used for locomotion and directing food to the mouth region.

If they are to survive, the functioning of the **cilia** needs to be carefully coordinated, otherwise, for example, there would be no effective movement towards a food source. How then does a single cell achieve this control? All the cilia are spontaneously active, but one cilia, termed the **pacemaker**, beats at a faster rate and the other cilia beat to the frequency of this cilia. The pacemaker drives all the other cilia by virtue of the fact that they are all coupled with each other through the aqueous environment in which they live. The direction of beat is, in part, controlled by the potential across the membrane of *Paramecium*. For example, when meeting an obstacle in its path, it reverses its direction and then continues in a forward direction missing the obstacle. The collision between the animal and obstacle is believed to result in the opening of Ca^{2+} channels, which leads to depolarization and causes the direction of beating of the cilia to reverse.

2.5.2 Nerve nets

Nerve nets are the simplest examples of nervous systems 'proper'. They are found in such animals as the corals, comb jellies and jelly fish. The arrangement of nerve nets in jellyfish is shown in *Figure 2.14*. Nerve nets simply represent a meshwork of neurons running around the body of these animals. Some of these nets represent specific neural pathways

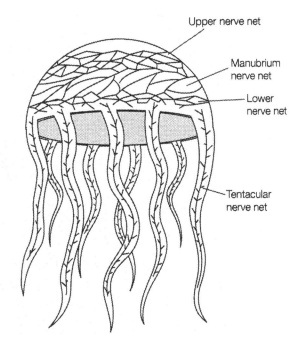

Upper nerve net

Manubrium nerve net

Lower nerve net

Tentacular nerve net

Figure 2.14. The nerve net of the jellyfish consists of an apparently random network of neurons which run through the bell and tentacles. Parts of the nerve net may function with some degree of independence from other parts, suggesting some specialization of certain neurons and neural pathways.

connected to a particular function, or a function of a particular region of the body. Others serve a more general purpose. Nerve nets in the 'bell' of the jellyfish coordinate movement of the animal. Others in the tentacles have a sensory function which may be important in the detection of food.

2.5.3 Nerve cords and cephalization

A major evolutionary trend in animals in phyla above the cnidarians was the development of nerve cords and **cephalization**, the development of a head. Such animals display bilateral symmetry, which means they have a right and a left side and a distinct front and rear end. Nerve cords are tracts of neurons collected together through which virtually all information passes. The simplest animals displaying a nerve cord are the flatworms, shown in *Figure 2.15*. However, even in this group it is possible to see a wide variety, with the simplest flatworms having a nervous system which has the appearance of a nerve net, whilst more advanced forms have a much better defined nervous system with a variable number of nerve cords. The next major development was the brain. A **brain** can be simply defined as a collection of neurons at the front end of an animal. Associated with cephalization and the development of the brain was the concentration of sensory structures at the front end of the animal. This was obviously a development favored by natural selection since it allowed animals to know where they were going as opposed to where they had just been. The appearance of the brain is once again seen first in the flat worms. As we move up through the phyla, the brain becomes increasingly complicated. For example, as

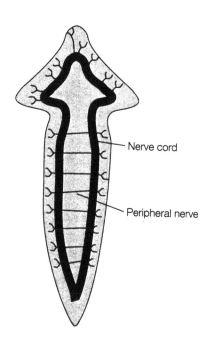

Figure 2.15. The nervous system of a flatworm showing the longitudinal nerve cords. The precise number of cords varies between species.

Nerve cord

Peripheral nerve

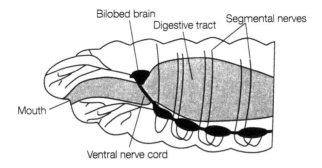

Figure 2.16. The annelid nervous system. From Withers, P., *Comparative Animal Physiology*, 1992, Saunders College Publishing.

shown in *Figure 2.16*, the annelid brain is a bilobed structure with a double nerve cord. Each segment of the annelid also has a pair of ganglia (collections of neural cell bodies and dendrites), which are generally motor in nature. The brain of a typical arthropod, such as locust (*Figure 2.17*), displays even greater complexity. It is generally a three lobed structure, each lobe now dealing with a specific type of sensory input. Furthermore, the individual **ganglia** of segments have now become fused together. In insects, for example, thoracic ganglia have developed which control the legs and wings.

Vertebrates have taken the development of the nervous system several stages further. Their nervous system consists of a brain and a single nerve cord, which together constitute the central nervous system. This can be contrasted with the peripheral nervous system which consists

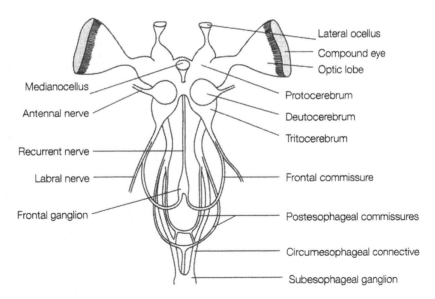

Figure 2.17. Anterior view of the brain and nervous system of the locust. Redrawn from Chapman, R.F., *The Insects*, 1982 with permission from Harvard University Press.

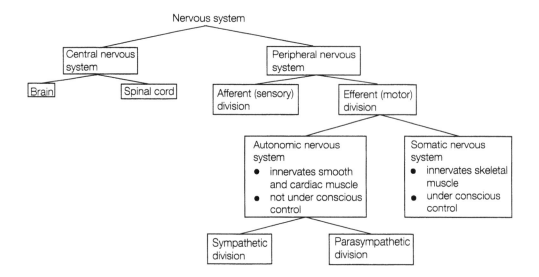

Figure 2.18. The organization of the vertebrate nervous system.

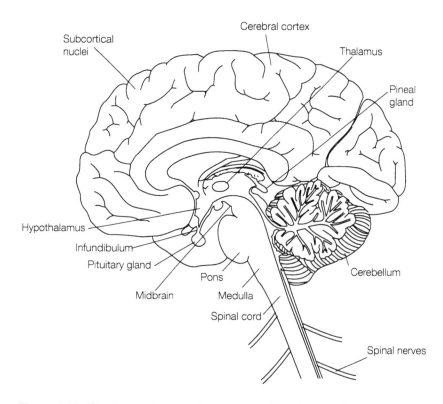

Figure 2.19. The human brain and spinal cord. The three main embryological regions are the rhombencephalon (cerebellum, pons and medulla), the mesencephalon (midbrain) and the prosencephalon (thalamus, hypothalamus and cerebrum).

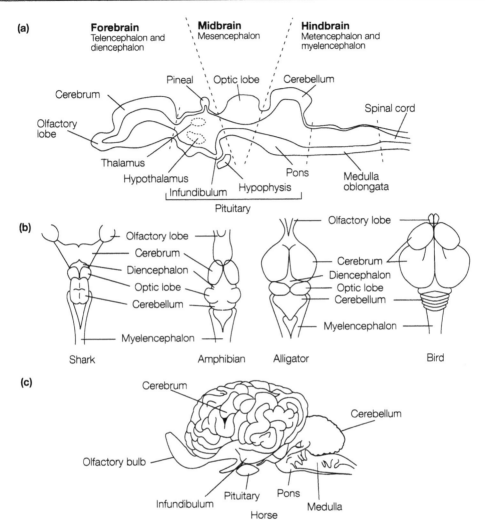

Figure 2.20. The brain structures of different vertebrates. (a) A median section through a typical fish brain; (b) dorsal views of the brains of four different vertebrates; (c) a typical mammalian brain, in this case the horse. Note the increased size of the cerebrum in the horse compared with the more primitive vertebrates. (a) and (b) from Weisz, P., *The Science of Zoology* 2nd edn, 1973, published by McGraw-Hill; (c) from Romer, A.S. and Parsons, T.S., *The Vertebrate Body*, 1986, published by Saunders College Publishing.

of pairs of a variable number of nerves originating from the central nervous system (*Figure 2.18*). The vertebrate brain can be divided into three main regions (*Figure 2.19*): the **prosencephalon**, the **mesencephalon** and the **rhombencephalon**. Furthermore, there have been considerable differences in the development of the brain among the vertebrates. For example, in primates, compared with lower vertebrates, there has been considerable development of the prosencephalon, particularly the cerebral cortex. This is shown in *Figure 2.20*, which demonstrates the organization of the vertebrate brain. Further details of the organization and functions of the vertebrate brain are beyond the scope of this text.

Further reading

Nicholls, J. G., Martin, A. R. and Wallace, B. G. (1992) _From Neuron to Brain_, 3rd edn. Sinauer Associates Inc., Sunderland, MA.

Sheperd, G. M. (1994) _Neurobiology_, 3rd edn. Oxford University Press, Oxford.

Underwood, P. N. R. (1977) _Nervous Systems_. Edward Arnold, London.

Receptors and effectors

3.1 Introduction

So far, the functioning of individual nerve cells and how they can be used collectively to form nervous systems has been described. Collectively, nervous systems provide information about what is happening in the immediate environment of an animal. This may either be the internal environment, e.g. the concentrations of particular ions in the body fluids, or the external environment, e.g. the salinity of the water in which a particular aquatic animal finds itself living. Equally, they must also be in a position to respond to this information regarding the state of their environment – they must be able to initiate appropriate biological responses. This chapter deals with the mechanisms by which animals collect information from their environment and how they are able to respond to such information.

3.2 Sensory neurophysiology

In order for animals to monitor change in their immediate environments, it is necessary that they are equipped with appropriate **sensory receptors** which allow them to 'pick up' the vast amount of information that bombards them. Information in this sense would include aspects of the environment, such as temperature, chemical composition, the presence of light and so on. The structure of sensory receptors varies widely. At the simplest level, they can simply be dendritic endings of neurons, usually **unmyelinated neurons**, as in the case of nociceptors or pain receptors. At their most complex, they involve the use of specialized nonneural receptors which pass 'information' onto neurons, as in the case of 'hair cells', which are the sensory receptors in the auditory and **vestibular systems** of vertebrates. The structure of a variety of sensory receptors is shown in *Figure 3.1*.

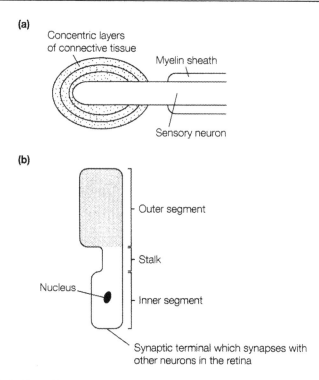

Figure 3.1. Two diverse types of sensory structure. (a) Pacinian corpuscle, a mechanoreceptor sensitive to pressure. (b) The vertebrate photoreceptor. The outer segment contains photosensitive molecules which capture photons of light.

3.3 Classification of sensory receptors

There are two ways of classifying sensory receptors. The first of these is based on the type of stimulus, i.e. their modality. There are six basic types of sensory receptor in this classification: **chemoreceptors, thermoreceptors, mechanoreceptors, photoreceptors, magnetoreceptors, electroreceptors**. (Specific examples of these will be discussed later.)

The second means by which sensory receptors may be classified is by their location. Thus, there is a group of sensory receptors which monitor internal conditions, known as **interoceptors**, e.g. those monitoring ionic composition of body fluids, and another group which monitor external conditions, the **exteroceptors**, e.g. those concerned with reception of sound waves.

3.4 Sensory receptor function

The overall job of sensory receptors is to act as transducers. That is, they convert (transduce) energy from one form, e.g. light, temperature, into electrical activity, ultimately action potentials. The process is summarized in *Figure 3.2.* The first step in the process is a change in the ionic permeability of the membrane of the sensory receptor, thus altering its ionic conductance. This can be achieved in several ways. For example, photoreceptors, i.e. receptors which are responsive to light, trigger a series of reactions which alter the state of opening of ion channels. For example, in invertebrate photoreceptors, Na^+ ion channels open causing a depolarizing response; in vertebrates, the opposite happens – the Na^+ channels are already open and light causes them to close as the photoreceptors hyperpolarize. Mechanoreceptors are sensory receptors which are sensitive to changes in pressure. These receptors may be sensitive to touch or pressure located on the general body surface or within deeper body tissues, or receptors sensitive to sound waves present in the external medium. Mechanoreceptors have ion channels which can be opened or closed by mechanical deformation, e.g. increased pressure. This physical deformation of the membrane causes specific ion channels to open. Imagine such deformation resulting in the opening of Na^+ channels. This means that Na^+ is free to enter the cell and can therefore produce a depolarizing response. This initial depolarizing change in membrane potential is converted into a local potential of the sort described in Chapter 2. The greater the stimulus, the greater the local potential. For the animal to perceive any sensation, the local potential must be converted into action potentials which are the 'currency' of the nervous system. How this is done depends upon whether the receptor is a free nerve ending or whether it is a specialized cell.

In the case of free nerve endings, such as the Pacinian corpuscle (an example of a vertebrate pressure detector), the local potential, which

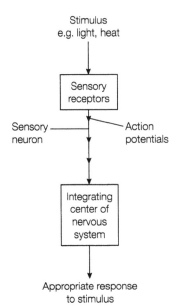

Figure 3.2. The transduction of environmental stimuli into action potentials is the role of sensory receptors. The sensory receptor may be a specialized dendritic region of the sensory neruon or it may be a specialized cell.

in this situation is called a **receptor potential**, is transmitted by local currents to the first node of Ranvier. If the change in membrane potential is sufficient to bring the cell membrane to threshold at the first node, an action potential is generated. This is depicted in *Figure 3.3*. In the case of sensory receptors which are specialized cells, the local potential, which in this situation is called a **generator potential**, results in the release of neurotransmitter. The neurotransmitter is then able to influence the electrical activity of the neuron associated with the sensory receptor. A good example is in vertebrate hearing, where sound waves cause the hair cells of the inner ear to depolarize and release neurotransmitter onto the auditory neurons. The effect of this is to generate action potentials in the auditory neurons, which are conveyed back to the region of the brain that deals with auditory information. It is at this site that the action potentials are translated into sound and where the perception of sound occurs.

The relationship between the intensity of the stimulus and the response generated is a relatively complex one. In many cases the relationship is not a simple linear one. Many sensory receptors, when subjected to a maintained stimulus, respond by decreasing the frequency of action potential production. This phenomenon is called **adaptation**. Some receptors adapt very quickly (this is called phasic adaptation), whilst others adapt very slowly (tonic adaptation). Sensory receptors which display phasic adaptation include those which deal with pressure and touch, for example. This is easily demonstrated – the perception of sustained pressure on your own skin by the weight of clothes is soon lost. In contrast to this, sensory receptors which display tonic adaptation include those which mediate the sensation of pain. Obviously, since pain is a protective response, it would be disadvantageous to animals to 'switch off' the activity of pain receptors in the presence of continual pain and its cause.

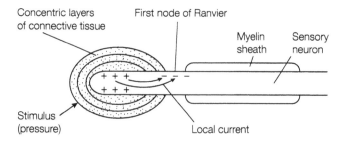

Figure 3.3. Generation of local potentials in the Pacinian corpuscle. Pressure applied to the end terminal results in depolarization (see text for explanation).

3.5 Sensory reception

Listed earlier were the types of sensory information that could be detected by animals. What follows is a brief elaboration, detailing what types of sensation within these general groupings can be detected by animals.

3.5.1 Chemoreception

Chemoreception is the detection of specific chemicals by a sensory receptor. Interaction of the chemical with its receptor results in the generation of receptor potentials, and, consequently, action potentials. Under the heading of chemoreception are included the senses of taste and smell, together with less obvious aspects such as the monitoring of O_2 and CO_2 levels in body fluids.

Chemoreception is a sense which has been demonstrated to exist in both invertebrates and vertebrates. Indeed, the chemoreceptor senses of invertebrates have quite often provided useful models for studying the physiological processes involved in the functioning of this sensory modality in higher animals. For example, the antennae of insects are easy structures to utilize in neurophysiological studies. Their ease of access means they can easily be stimulated and their response to such stimulation is easy to record. Taste is a common sensation throughout the animal kingdom. In particular, the ability to taste bitter substances is almost universal in the animal kingdom, as it serves a protective purpose by warning against potentially toxic substances. The role of smell is also important in many animals, **pheromones** are a good example of this. These are volatile chemicals released into the atmosphere which are used as signals to other animals. For example, the female silk worm moth, *Bombyx mori*, releases a chemical called bombykol which is used to attract a male mate. The male has receptors for bombykol in its antennae. It is thought that a single molecule of bombykol in 10^{15} molecules of air will evoke a response in the male to begin searching for a female. This obvious and directed movement towards a chemical substance is called **chemotaxis**. Similar mechanisms also exist in the vertebrates. For example, the sense of **olfaction** is thought to be involved in the process by which some fish, for example, salmon, return to their home waters to reproduce. In snakes and some lizards, there is a specialized olfactory structure called Jacobson's organ, which is located in a little outgrowth of the nasal cavity. The characteristic trait of many snakes is the continual movement of the tongue back and forth into the mouth. What they are actually doing is transferring environmental air samples onto their Jacobson's organ for analysis – essentially, they are tasting the air. Humans have the ability to distinguish between many thousands of smells. However, just as there are only three primary colors from which all other colors are produced, there are also thought to be only a small number of 'primary' odors.

3.5.2 Mechanoreception

Mechanoreception is the measurement of force (pressure) and displacement. The receptors which monitor this can vary from simple dendritic endings of neurons, e.g. receptors in the skin which monitor pressure, through to highly complicated, specialized receptors, e.g. the inner ear of vertebrates. Although a common sensation, the actual physiological process whereby signals are transduced is far from clear. The simplest mechanism proposed is that stretch-activated ion channels exist in the membrane of the receptor components. At rest, the ion channels are closed. However, the application of pressure produces a conformational change in the protein that constitutes the ion channel, and this leads to depolarization and the generation of local currents. This is shown in *Figure 3.4*.

Mechanoreception is seen to be present in both invertebrate and vertebrate animals. In invertebrates, it has been shown that receptors exist for pressure, sound and movement detection. For example, insects have surface receptors which provide information about wind direction, the orientation of the body in space, velocity of movement and sound. In a similar manner, the mechanoreceptors of vertebrates are equally varied. They range from receptors which can monitor the length of a muscle (specialized muscle fibers called muscle spindles) through to the organs of hearing and balance (inner ear structures). In the case of the latter, the sensory receptors are ciliated (*Figure 3.5*), and it is the movement of the cilia which is responsible for the production of action potentials. Perhaps the simplest organization of this kind is the **lateral line** system of fish. The lateral line (*Figure 3.6*) provides information about the movement of the animal itself, as well as information about other movements nearby. Further evolutionary development of the lateral line system gave rise to the inner ear of vertebrates. It is beyond the scope of the present text to discuss auditory mechanisms in detail. However, many variations of the basic hearing process have developed in animals. Microchiropteran bats, for example, depend on echolocation – they hear the echoes of sounds that they produce themselves which are reflected back by other objects. Given that echolocation is used by animals active at night, it can be used for both prey detection and the avoidance of objects. Echolocation is also used by other animals, such as cave-dwelling birds, whales and dolphins.

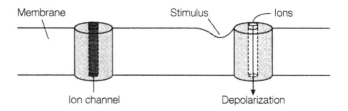

Figure 3.4. Schematic diagram of a stretch-activated ion channel.

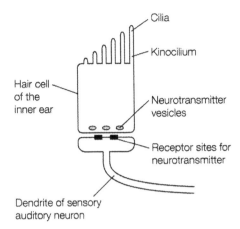

Figure 3.5. Displacement of the cilia in the inner ear by sound waves results in the release of neurotransmitter. This causes the sensory auditory neuron to generate action potentials.

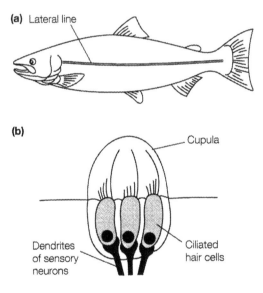

Figure 3.6. (a) The lateral line in fish. (b) Along the lateral line are many hair cells, the cilia of which are covered by a gelatinous mass called the cupola. Displacement of these provide the fish with information about its own and other movement.

3.5.3 *Thermoreception*

Thermoreception is the detection of heat and cold and changes in temperature in the environment (either external or internal). It is an important parameter to be able to measure since it can have a profound effect upon the functioning of the animal. For example, consider the consequences of increased temperature upon protein structure and function and the effects this would have on metabolic processes. At increased temperatures, proteins and, therefore, enzymes would cease to function and the necessary metabolic reactions essential for life would be unable to occur. Thermoreception occurs throughout the animal kingdom. Virtually all animals have an optimum temperature within their environment; for example, under experimental conditions, parasitic worms will move to regions of higher temperature rather than move to regions of lower temperature. The location of thermoreceptors varies between animals. Insects, for example have thermoreceptors on their antennae and legs for monitoring air temperature and ground temperature, respectively. Mammals, on the other hand, have peripheral and core thermoreceptors located in the skin and hypothalamus, respectively. This is essential if a constant core body temperature is to be maintained, in this case 37°C. The system would not work if there were only central receptors. Consider what would happen if there was a dramatic rise in the temperature of the external environment. By the time such an increase had been transmitted through to core thermoreceptors, serious heat-induced damage may have already occurred to the animal in the periphery.

3.5.4 Photoreception

The ability to detect light is an almost universal phenomenon across the animal kingdom. Although some organisms, e.g. *Amoeba*, may be able to detect the presence of light without the aid of any specialized structures (how this is achieved is far from clear), it is more usual for an animal to possess a specialized photoreceptor. This may vary from the light-sensitive cytoplasmic region (eye-spot) of *Euglena* to the complicated neural organization of the vertebrate eye. In all cases, though, the operation of photoreceptors is essentially the same. Highly folded photoreceptor cells contain pigments, the most common of which is **rhodopsin**, which are chemically changed in the presence of light through a series of intricate reactions that result in alteration in the membrane potential of the receptor. The physiological processes which are occurring at photoreceptors, and the way that neural information generated from them is dealt with by the nervous system, is far from fully understood. Flatworms have so-called cup-shaped eyes (*Figure 3.7*) which face in opposite directions and are used essentially to obtain directional information. The animal will tend to move to darker regions, thereby minimizing the risk of attack by predators. The arthropods (insects, crustaceans and spiders) have **compound eyes** (*Figure 3.8*). Each compound eye is made from many smaller optical units called ommatidia. The quality of the representation of the world varies depending on the number of ommatidia – the greater the number, the better the definition of the visual experience. A simple analogy here would be the printed output from a dot-matrix printer. The greater the number of dots which make up the print, the greater the clarity of the image. Vertebrates and cephalopod molluscs have developed **vesicular eyes** (*Figure 3.9*) which have the ability to form

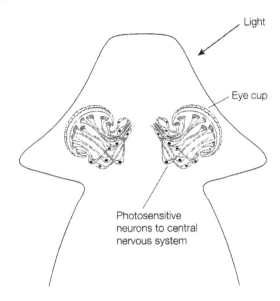

Light

Eye cup

Photosensitive
neurons to central
nervous system

Figure 3.7. The cup-shaped eyes of a planarian flatworm. The eye cup is a layer of pigment cells.

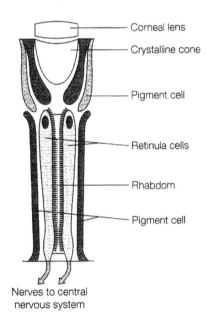

Corneal lens

Crystalline cone

Pigment cell

Retinula cells

Rhabdom

Pigment cell

Nerves to central
nervous system

Figure 3.8. An ommatidium, a single element of an insect compound eye. Hundreds of ommatidia make up the compound eye. From Miller, S.A. and Harley, J.P., *Zoology* 3rd edn, 1996, Wm. C. Brown Communications Inc. With permission of The McGraw-Hill Companies.

high definition images of the external world. The iris is able to alter the amount of light entering the eye and the lens focuses the light on the retina, the layer of photoreceptors at the back of the eye. The lens operates differently in different vertebrates. For example, fish, amphibians and reptiles move the lens nearer or further away from

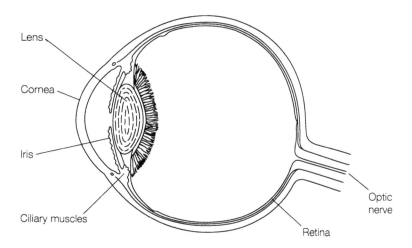

Figure 3.9. Section through a vertebrate eye. Movement of the ciliary muscles alters the shape of the lens. This focuses light rays on the back of the retina where the photoreceptors are located.

the retina in order to focus on a particular object. Birds and mammals on the other hand alter the shape of their lens so that light is focused on the retina. The organization of the vertebrate retina is exceedingly complex and is shown in *Figure 3.10*.

3.5.5 Electroreception

Some animals, particularly sharks, rays and catfish, have the ability to detect the minute electric fields generated by other animals. Such electric fields are generated by muscular activity, for example. This represents one way by which prey may be detected. The electroreceptors, the best known example of which is the ampulla of Lorenzini (in the sharks and rays), are located mainly in the head and along the lateral line. The dogfish, for example, can detect electric fields as small as 10 nV per cm. Some fish, the so-called electric fish, produce a continuous electric field around them. Disruptions to these fields by other fish or inanimate objects in their environment are detected by the electroreceptors. Living and nonliving objects are distinguished by the way they disrupt the electric field.

3.5.6 Magnetoreception

Many animals display the ability to orientate to the Earth's magnetic field. Such an ability allows animals to distinguish a north-south axis, which is potentially useful as a navigation aid. However, it is a sensory modality which is not widespread throughout the animal kingdom. In invertebrates, it is known that honey bees utilize the Earth's magnetic field in order to communicate. When bees arrive at their hive having

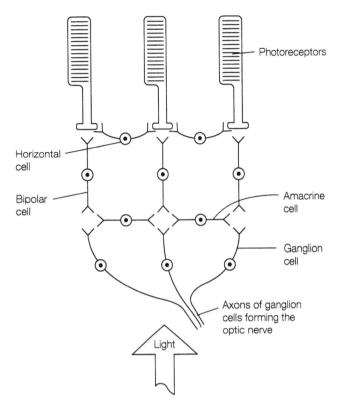

Figure 3.10. A simplified diagram of the neural organization of the vertebrate retina. The complexity of this organization allows some of the early processing of visual information to occur here.

found a new food source, they pass on information about the direction of the source to other bees by performing a waggle dance. In this way they communicate the angle between the sun and the food source, together with whether the source is away from or towards the sun. In fish, the electroreceptors are thought to be sensitive enough to detect a magnetic field. Homing birds are also thought to use magnetic fields as one of many cues that allow them to return home. The actual mechanism that allows for magnetoreception is far from clear. Some sharks, honeybees and birds have been shown to contain the magnetic substance magnetite. One theory is that the magnetite allows these animals to orientate to the Earth's geomagnetic field. This provides them with directional clues and enables them to find their way around. It is interesting to note that bacteria have also been found which contain magnetite, although the implications of this finding are unclear.

3.6 Effectors – the responses to sensory information

Having gathered various types of sensory information, animals have to take appropriate action based upon the information gathered.

Effectors may be defined as anything capable of producing a biological response. This can range from muscles producing movement – an overt biological response – to endocrine glands secreting hormones to alter some aspect of metabolism, e.g. insulin reducing blood sugar levels. Many types of response are effected in response to sensory stimuli, some of which will be briefly mentioned now.

Some animals possess the ability to change the color of their skin. Such changes occur for a variety of reasons including camouflage and communication to other animals. Such signals may be to animals of the same species, for example, to obtain a mate, or to animals of other species, for example, to deter predators. The color change occurs because the skin of the animal contains pigment containing cells called **chromatophores**. Animals which possess such cells include squids, octopods, some flat fish (e.g. flounder), chameleons, frogs and snakes. The responses of chromatophores are shown in *Figure 3.11*. Chromatophores are under neural or endocrine control, or in some cases both. They have the ability to alter the dispersal of pigment within the cell in minutes or seconds, thus altering the overall appearance of the animal. The mechanism by which this occurs differs from species to species. In cephalopods (e.g. squid and octopus), each chromatophore is bounded by muscle cells and the contraction and relaxation of these muscles alter the dispersion of the colored pigment. When the muscles contract, the chromatophore 'enlarges' and the pigment disperses. When the muscles relax, the chromatophore shrinks and the pigment becomes concentrated. Contraction and relaxation of these muscle cells is under neural control. In contrast, the chromatophores of amphibians function by simple dispersion of the pigment within the chromatophore. In many cases, this is under endocrine control.

Virtually all animals possess a system of glands whose secretions result in a variety of biological responses and which are activated under appropriate conditions. Many of these will be discussed in Chapter 4. However, there are many other glands which serve more specialized functions. Some of these have already been mentioned, e.g. glands

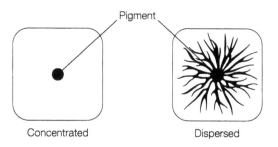

Figure 3.11. The responses of a chromatophore. The pigment within the cell may be either concentrated or dispersed.

which release attractant pheromones. Other examples of such glandular secretions serve a more aggressive role. For example, skunks release a foul-smelling substance when attacked. The bombardier beetle, *Brachinus*, sprays attackers with a fluid which has a temperature of 100°C. The tip of the abdomen of this beetle contains two separate chambers. One chamber contains the reactants necessary for the production of this hot fluid; the second chamber contains the necessary enzymes required for the reaction to proceed. The release of reactants into this second chamber allows the production of this hot fluid to occur and its release at potential predators. In addition to the two examples given above, there are many more subtle responses which are not observed. For example, the release of **insulin** into the bloodstream after a meal in order to maintain appropriate blood sugar levels is a response to disturbance of blood sugar levels.

However, as previously stated, perhaps the most obvious response that animals make is that of movement.

3.7 Intracellular movement

Before discussing movement of whole animals, it is worth considering the movement of substances within cells. All animal cells are dynamic structures, with organelles and other substances continually moving. However, it is worth remembering that order is superimposed upon this dynamic structure. This order comes from the cytoskeleton, which, as the name implies, is the internal skeleton of all animal cells. The **cytoskeleton**, which contributes to the shape of the cell, its movement, and movement within the cell is composed of three types of fibre: microfilaments, microtubules and intermediate filaments. **Microtubules** are polymers of a protein called tubulin and these elements make up the cilia and flagella which can be found in animal cells. **Microfilaments** are also proteinacious structures, formed from proteins called actins, sometimes together with other protein molecules, such as myosin. As will be seen later, these structures are important in making cells shorten – the type of response which occurs when muscles contract. The final group are the **intermediate filaments**. These structures play a less dynamic role in cell movement and are probably important in stabilizing cell structure and resisting movement.

In virtually all cells, the cytoplasm of the cell is continually moving, a phenomenon called cytoplasmic streaming. It should be remembered that this is a directed movement, i.e. it does not occur randomly. A good example of this cytoplasmic movement is seen in **axonal transport**. Here, microtubules run along the entire length of the axon from the cell body to the axon terminals. Movement of substances occurs in both directions, from the cell body to the axon terminals and vice versa. Some neurotransmitters are transported from the cell body to the axon, known as anterograde transport. This may reach speeds of up to 40 cm day^{-1}. Flow in the opposite direction, retrograde flow, carries material from the axon terminal back to the cell body for reprocessing and occurs at a much slower rate of about 8 cm day^{-1}.

3.8 Amoeboid movement

Amoeboid movement is a characteristic movement of both unicellular animals, such as _Amoeba_, and also the cells of multicellular animals. For example, white blood cells in vertebrates use amoeboid movement to leave the circulation and enter the tissues where they may become involved in inflammatory reactions, and during the early development of animals, many cells move to their final destination by such movement. Amoeboid movement by single-celled animals involves the animal extending the cell to form a pseudopodium (_Figure 3.12_). The precise mechanism behind this movement is unclear, but it is thought to involve changes in the physical nature of the cell cytoplasm. The general cytoplasm of the cell is liquid in its nature and is known as **plasmasol**. However, around the periphery of the cell it is much more viscous in nature, and here it is called **plasmagel**. Pseudopodia are formed by the breakdown of the plasmagel cytoplasm in a particular region of the cell membrane. Positive pressure is generated in the main part of the cell by the interaction of microfilaments. There are two types of microfilament which form a network in the cytoplasm – these are actin and myosin. Actin and myosin interact with each other in a similar manner to the interaction of actin and myosin in muscle contraction, described in Section 3.9.2. This interaction forces the plasmasol past the breakdown of the plasmagel, which causes the cell well to bulge outwards at that point to form a pseudopodium. As the plasmasol enters the pseudopodium it changes into plasmagel and the pseudopodium is thus prevented from forming any further.

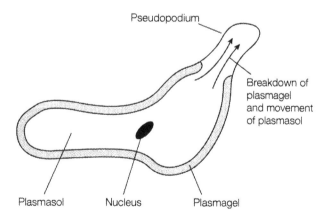

Figure 3.12. Amoeboid movement is achieved by the breakdown of the plasmagel at the extending region and the flow of plasmasol into this region. Plasmasol then changes to plasmagel.

3.9 Muscle and movement

3.9.1 Muscle types

Muscle may be roughly divided into two types: **striated muscle** and **smooth muscle**. Striated muscle itself may be further subdivided into **skeletal muscle** and **cardiac muscle**.

Skeletal muscle is under voluntary control. The fine structure of skeletal muscle will be described later. Cardiac muscle, as the name suggests is muscle which makes up the heart. One of its characteristic features is the presence of junctions between individual cells called **intercalated discs**, sometimes called gap junctions. This is shown in *Figure 3.13*. Intercalated discs represent regions of low electrical resistance between the individual muscle cells which constitute the heart. They are important in ensuring that action potentials are rapidly transmitted across the heart, so that all the muscle cells contract in unison and the heart beats synchronously.

Smooth muscle is so called because it lacks the striations that give striated muscle its characteristic appearance. On the basis of its neural innervation, it is possible to distinguish two different types of smooth muscle, visceral, or single-unit smooth muscle, and multi-unit smooth muscle. Single-unit smooth muscle is characterized by the appearance of gap junctions – cellular junctions rather like the inter-calated discs found in cardiac muscle – between individual cells. These junctions allow action potentials to be transmitted between groups of individual cells very rapidly. Thus, a wave of contraction may appear to pass over such smooth muscle as an action potential is transmitted from one cell to the next. Multi-unit smooth muscle, on the other hand, requires each smooth muscle cell to be innervated in order to generate a contractile response.

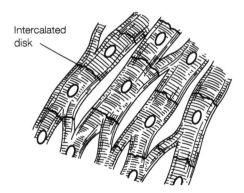

Intercalated disk

Figure 3.13. Cardiac muscle has a striated appearance and extensive branching. The region where one cell joins the next is called an intercalated disc.

In the higher vertebrates, smooth muscle surrounds internal organs, such as the gastrointestinal tract and the airways, and, generally, it is not under voluntary control. However, it is important not to make such generalizations across the animal kingdom as a whole. For example, the muscle associated with the gut in arthropods is striated in appearance, but it is not under voluntary control, unlike striated muscle in vertebrates.

3.9.2 Contraction of vertebrate skeletal muscle

The detailed structure of skeletal muscle is shown in *Figure 3.14*. Many of the earliest studies investigating the mechanism of muscle contraction were performed using skeletal muscle. Muscles are made up of many muscle fibers, equivalent to muscle cells. The cell membrane which surrounds the muscle cell is called the **sarcolemma** and it has a number of invaginations into the cell called transverse (T) tubules. The cytoplasm of muscle cells is called the **sarcoplasm** whilst the endoplasmic reticulum of muscle cells is called the **sarcoplasmic reticulum**. Muscle fibres in turn are composed of **myofibrils**. The myofibrils are seen to be banded (or striated, hence the name striated muscle) in appearance. Some of the bands (I bands) appear to be lighter than other bands (A bands). This is because I bands contain only thin filaments (actin), whereas A bands are regions where thin filaments overlap with thick filaments (myosin). Hence, the A regions appear darker. Each I band is split into two by the presence of the Z line, a region where the thin filaments of adjacent sarcomeres are connected by a protein called α-actinin. The sarcomere is the basic unit of contraction and extends from one Z line to the next. In turn, the myofibrils (and therefore the sarcomeres) are composed of **myofilaments**. There are two types of myofilaments present in striated muscle. The first is a group of thick filaments, which are composed of the protein **myosin**. Myosin consists of two polypeptide chains which wrap around each other to form a coil. Several of these coils group together in a bundle with their myosin heads facing outwards to form a thick filament. In addition, there is a group of thin filaments, which consist mainly of the protein **actin**, together with two other proteins called **tropomyosin** and **troponin**. These thin filaments are formed from two actin molecules coiled together with two tropomyosin molecules wrapped around them. Pairs of troponin molecules occur at intervals along the actin molecule. Tropomyosin and troponin are referred to as regulatory proteins. The structure of the myofilaments is shown in *Figure 3.15*. It is the overlap of the myofilaments in the myofibrils which gives skeletal muscle its characteristic banded appearance. Actin and myosin overlap in the A band. The region where there is no overlap is called the H band.

How, then, does contraction occur? The mechanism by which muscle contracts is known as the sliding filament model. With this model, the actin and myosin filaments slide past each other without physically shortening. At rest, when the muscle is relaxed, the process of contrac-

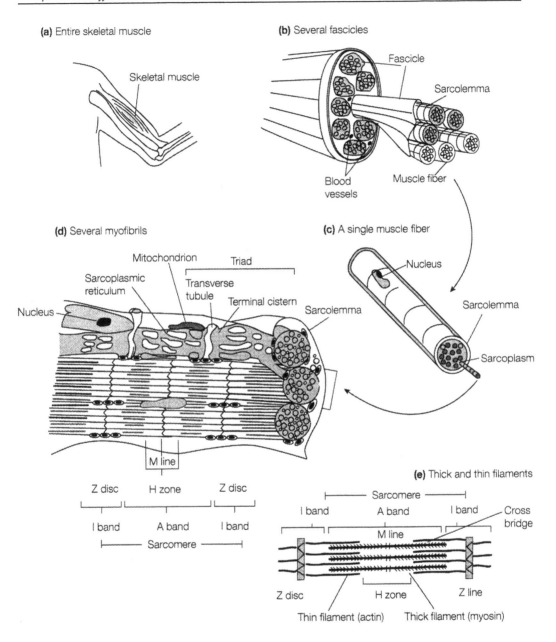

Figure 3.14. The organization of vertebrate skeletal muscle. A whole muscle is made of muscle fibers which are made up of myofibrils. The organization of the myofibrils gives skeletal muscle its characteristic striated appearance. Adapted from Tortora, G.J. and Grabowski, S.R., *Principles of Anatomy and Physiology*, 1996, published by Addison Wesley Longman.

tion is prevented by the regulatory proteins troponin and tropomyosin, which prevent the formation of cross-bridges that are necessary for contraction (see later). The result of the filaments sliding past each other is that the Z lines move closer to each other and the muscle shortens.

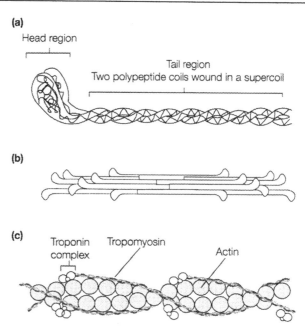

Figure 3.15. Structure of the thick and thin filaments in skeleton muscle.
(a) Myosin molecule; (b) thick (myosin) filament; (c) thin (actin) filament. From
Hickman, Roberts and Larson, *Integrated Principles of Zoology* 10th edn,
p. 643, 1996, Wm C. Brown Communications Inc.

The contraction of skeletal muscle is initiated by the release of a neurotransmitter from the nerve which innervates the muscle. In the case of
vertebrates, this is the release of acetylcholine from the axon terminals
of motor neurons. A single motor neuron and all the muscle fibers
which it innervates is termed a **motor unit**. A single neuron may innervate a single muscle fiber or anything up to 100 muscle fibers. In terms
of overall muscular activity, the fewer the number of muscle fibers in
an individual motor unit, the more precise and fine is the movement
of that particular muscle. Equally important are the total number of
motor units which serve a particular muscle. In primates, for example,
the muscles of the hands and fingers are composed of many motor
units. Thus, very intricate movements of these appendages may be
performed. This contrasts with the muscles of, say, the trunk. These
are composed of fewer, larger motor units and the movements achieved
by such muscles are much broader.

The interaction of the neurotransmitter with the muscle results in depolarization of the muscle cell. This initial depolarization spreads across
the entire muscle cell membrane including transmission down the T
tubule system deep into the myofilaments. The effect of depolarization
is to release Ca^{2+} from the sarcoplasmic reticulum where it is normally
stored at rest. The release of Ca^{2+} into the sarcoplasm is the trigger
for contraction. Ca^{2+} released into the sarcoplasm binds to troponin,
inducing a conformational change in the shape of the troponin mole-

cule. This in turn causes a change in the orientation of the tropomyosin molecule, which moves exposing a binding site for myosin on the actin filament. The exposure of the myosin binding site on the actin filament allows the two myofilaments to bind to each other – this is known as cross-bridge formation. The head of the myosin filament which binds to actin also possesses ATPase activity, and it converts ATP to ADP and inorganic phosphate (P_i). In doing this, energy is released, which is stored in the head region. The ATPase activity described above occurs before cross-bridge formation takes place. Upon the formation of cross-bridges, the energy stored in the myosin is released. The effect of this is to pull the thin filaments towards each other, thus shortening the length of the sarcomere and causing the muscle to contract. During cross-bridge formation, the ADP and P_i, which are bound to the myosin head region, become detached. Their removal allows the bonding of another molecule of ATP. This causes the cross-bridge binding between actin and myosin to cease. The molecule of ATP is broken down to ADP and P_i, leading to the reformation of cross-bridges which pull the thin filaments further towards each other. The cycle continues and the microfilaments are pulled past each other causing the muscle to contract. The whole process is terminated when repolarization of the muscle cell occurs. During repolarization, Ca^{2+} is removed from the muscle cell and is stored in the sarcoplasmic reticulum from where it was initially released. This allows the tropomyosin molecule to return to its original position on the actin molecule thereby blocking the myosin binding site. Given the role that troponin and tropomyosin play in muscle contraction, it is understood why they are called regulatory proteins. The process of muscular contraction is summarized in *Figure 3.16*.

Although the above description relates to striated muscle, much the same sort of process occurs in other types of muscle. However, there are also significant differences. For example, cardiac muscle cannot enter into the state of maintained contraction known as tetany because it has an extremely long refractory period of 250 ms. This time period is almost as long as it takes for cardiac muscle to contract and fully relax. Therefore, it is impossible for individual contractions to summate to give a sustained contraction. Remember that during the refractory period, a cell is unresponsive to further stimulation. Obviously, if this were allowed to happen, then the heart would cease to function as a pump. Another difference is that in smooth muscle, Ca^{2+} entry from the extracellular fluid is important in initiating the contractile response, in contrast to the role of internally released Ca^{2+} seen in skeletal muscle.

3.9.3 Types of vertebrate skeletal mucle

It is possible to distinguish two main types of vertebrate skeletal muscle – fast (or twitch) fibers and slow (or phasic) fibers. Fast fibers represent those muscles which are utilized for rapid bursts of activity. Such fibers possess few mitochondria, little myoglobin and a poor blood

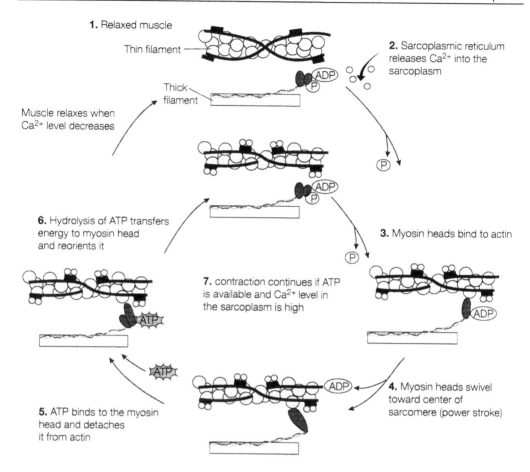

1. Relaxed muscle

Thin filament

Thick filament

Muscle relaxes when Ca^{2+} level decreases

2. Sarcoplasmic reticulum releases Ca^{2+} into the sarcoplasm

6. Hydrolysis of ATP transfers energy to myosin head and reorients it

3. Myosin heads bind to actin

7. contraction continues if ATP is available and Ca^{2+} level in the sarcoplasm is high

5. ATP binds to the myosin head and detaches it from actin

4. Myosin heads swivel toward center of sarcomere (power stroke)

Figure 3.16. The mechanism of muscle contraction. The myosin heads bind to actin in the thin filament and then detach, pulling the thin filament past the thick filament. The cycle continues and muscle contracts. The process requires ATP and Ca^{2+} (see text for explanation). Adapted from Tortora, G.J. and Grabowski, S.R., *Principles of Anatomy and Physiology*, 1996, published by Addison Wesley Longman.

supply. Consequently, much of their metabolism is anaerobic. This type of muscle is found, for example, in frogs' legs and is used to help the animal to jump. Slow fibers, on the other hand, are rich in mitochondria and myoglobin and have a good blood supply. These muscles are used where contraction needs to be maintained for longer periods, such as the maintenance of posture in terrestrial vertebrates.

3.10 Skeletal systems

Muscles only do biological work when they contract; relaxation is a passive process. Therefore, muscles are usually found in antagonistic pairs, so that when one muscle group contracts and performs work the other relaxes. However, in order to perform useful work, such as movement, muscles need something to pull against. In vertebrates, it is the skeleton against which muscles pull, whilst in invertebrates, e.g. annelid worms, it is the substrate across or through which they are moving that muscles pull against.

3.10.1 Hydrostatic skeletons

Hydrostatic skeletons are found in soft bodied invertebrates, such as annelid worms. In some respects, they function in a similar manner to the amoeboid movement found in unicellular animals and other motile cells. Essentially, they consist of fluid enclosed in a body cavity which is surrounded by muscle. In some cases, hydrostatic skeletons have become modified for the purposes of locomotion, e.g. annelid worms. Annelid worms are segmented and have their body cavity surrounded by both **circular** and **longitudinal muscle** layers. Circular muscle is arranged around the circumference of the animal, whereas longitudinal muscle is oriented along the length of the animal's body. Contraction of discrete regions of the circular muscle results in some segments becoming narrower and more elongated, whilst contraction of the longitudinal layer results in some segments becoming shorter and wider. Movement in these animals is achieved by alternate contraction and relaxation of these muscle layers in different body segments (*Figure 3.17*). In order to maximize movement, some of the segments of the worm have bristles called setae protruding from them. These bristles anchor certain regions of the worm to the surface across which it is moving and prevent it from moving backwards in the opposite direction.

3.10.2 Exoskeletons

Exoskeletons are skeletons on the exterior of the body and are found in molluscs and arthropods. In the case of molluscs, e.g. mussels, clams

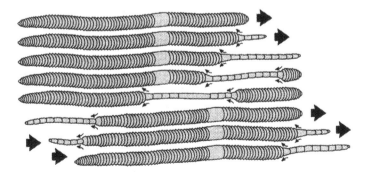

Figure 3.17. The use of the hydrostatic skeleton to achieve directional movement. Alternate contractions of longitudinal and circular muscle in the body wall pass down the length of the body. The animal is anchored by setae which prevent backward movement. From Hickman, C.P. and Roberts, L.S., *Biology of Animals*, 6th edn, 1994, Wm C. Brown Communications Inc.

and other bivalve molluscs, the exoskeleton takes the form of a shell and is primarily used for defensive purposes – the animal can retreat into the shell to escape predators, for example. In the arthropods, muscle is attached to the exoskeleton and, given that arthropods are segmented and jointed animals, contraction of the muscles allows the animal to move. The arthropod exoskeleton is a continuous structure, although there is tremendous variation in its flexibility; for example, flexibility is increased at joints, without which movement would be very limited. Overlying the exoskeleton, which is made of **chitin** (a complex polysaccharide), is the cuticle, a waxy substance which is an adaptation to minimize water loss. The biggest problem with such a skeleton is that it limits the animal's growth. Therefore, in order for arthropods to grow they must periodically shed their exoskeleton and grow a larger, new one. This leaves the animal in a vulnerable position. First, the new exoskeleton provides little defence until it has completely hardened, so the animal is open to increased attack from predators. Second, because the new exoskeleton is soft, movement is limited until it hardens, as muscle needs a rigid structure against which to contract.

3.10.3 Endoskeletons

Endoskeletons are internal skeletons. These are best developed in the vertebrates, although they are present in some invertebrates, e.g. the echinoderms, where they are composed of calcium salts. In vertebrates, the skeleton serves much the same function as the exoskeleton seen in other phyla, providing protection and a rigid framework against which muscles can contract to produce movement. In most vertebrates the skeleton is made of bone, which is principally made of calcium phosphate. However, there are some animals (e.g. the sharks and rays) which have a skeleton made from cartilage, which is principally composed of collagen – hence the term cartilaginous fishes by which these animals are known. The main advantage of the endoskeleton compared with the exoskeleton, is that endoskeletons grow continually with the animal, thus eliminating the need for moulting and the problems associated with this.

Endocrine function

4.1 Introduction

The endocrine or hormonal system (i.e. the use of body fluid-borne chemical messengers) together with the nervous system make up the control and coordinating systems of animals. However, there are major differences in the way in which control is achieved within the two systems. Firstly, the endocrine system works by transmitting chemical rather than electrical signals, although the nervous system utilizes chemical messengers at synapses. Secondly, the endocrine system has a much slower response time than the nervous system. An action potential is completed in 2–3 ms, but the action of hormones may take minutes or hours to be completed. Finally, endocrine action has a much longer duration of response. For example, reflexes in animals – fast pre-programmed responses of the nervous system – take a few milliseconds to be performed. Compare that with growth processes that are achieved utilizing the hormonal system that may take years to be completed. However, having stated that there are major differences between the two systems, it is becoming increasingly recognized that rather than working as two 'independent' systems, the nervous and endocrine systems work cooperatively to achieve a common goal. Indeed, some neurons will release neurotransmitters at their synapses that are then used to serve an endocrine function. Most animals have an endocrine system, and it controls many diverse physiological functions, e.g. metabolism, growth, reproduction, osmotic and ionic regulation, and so on.

4.2 Definition of endocrine systems

The classical idea of the endocrine system is that of cells, usually of a nonneural origin (although some neural tissue is considered to have an endocrine function), which secrete specific chemical messengers called **hormones.** The hormones are carried to their target organs (i.e. the organs where they exert their biological effect), usually some distance from their site of release, in the body fluids of the animal concerned. This idea is summarized in *Figure 4.1*. However, this classical view of endocrine organs and function has recently changed. For example, it is now known that some hormones do not need to enter the general circulatory system of animals in order for them to exert an effect. A good example of this is the role of **histamine** in controlling acid secretion in the vertebrate stomach, whereby various stimulatory factors converge on mast cells in the stomach (as well as parietal cells) leading to the release of histamine which, in turn, stimulates acid production. The organization of this is summarized in *Figure 4.2*. This type of 'local' hormone action is called **paracrine** control. In general terms, though, endocrine systems may be classified as one of two types. The first is the **neuroendocrine system**, also called the neuro-secretory system or neurosecretory cells. In this case, neurons are specialized for the synthesis, storage and release of **neurohormones** – in reality, this is the neurotransmitter of the neuron concerned. The neurohormone, instead of being released into a synapse, is released into the general circulation from where it travels to its target organ. The neuroendocrine system is found in all invertebrates and verte-brates. In mammals, for example, renal excretion of water is controlled by the secretion of **antidiuretic hormone** (ADH) released from neurons whose cell bodies lie in the hypothalamic region of the brain and whose axons extend down to the posterior **pituitary gland**. In some cases, the release of neurohormones into the general circulation may influence other endocrine organs which then exert some biological effect. For example, in crabs, moulting is controlled by the neurohormone moult inhibiting hormone (MIH), which in turn inhibits the activity of a second endocrine gland which produces a hormone that promotes moulting. The widespread presence of neuroendocrine control systems in both invertebrates and vertebrates suggests that they evolved earlier in evolution than the second type of endocrine system, the **classical**

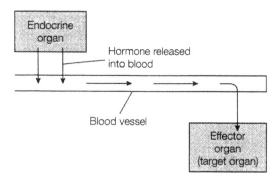

Figure 4.1. The typical organization of the endocrine system. Hormones are released directly into the blood and are carried to their effector organ, which is usually some distance from the site of release. In some cases, e.g. steroid hormones, the hormone is coupled to a transport molecule in the blood.

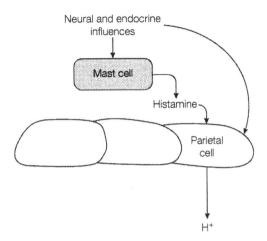

Figure 4.2. The role of histamine in acid secretion in the vertebrate stomach. Histamine is thought to act as a local hormone.

endocrine system. In this case, hormones are released from specialized, nonneural tissue directly into the body fluids. The absence of ducts to transport the hormone from the gland to the circulating body fluids (e.g. plasma, haemolymph) gives rise to the term **ductless gland**, an alternative term by which endocrine glands are sometimes described. This contrasts with **ducted glands** (e.g. salivary glands), where an anatomical duct leading from the gland to the body fluids is present. Classical endocrine glands are only found in the higher invertebrates (e.g. some molluscs) and the vertebrates. This suggests that the appearance of this system occurred after the development of the neuroendocrine system. The presence of neuroendocrine control systems and paracrine control blurs the typical definition and concept of endocrine control, and it may be more appropriate not to consider the two control systems of neural and endocrine control as being so clearly distinct from each other.

4.3 Identification of endocrine organs

It can be difficult to determine whether a particular structure in an animal serves an endocrine function. The fact that there are no unique anatomical markers that serve to identify endocrine from nonendocrine tissue is just one reason. In order to overcome this problem, criteria have been established by which candidate tissues and their secretions may be classified as true endocrine organs.

(i)   Removal of the candidate tissue or organ should produce deficiency symptoms. For example, if a tissue was suspected of producing a substance that maintained Na$^+$ levels in body fluids, then removal would result in disruption of Na$^+$ levels.

(ii)   Reimplantation of the candidate tissue or organ should result in the reversal or prevention of the associated deficiency symptoms. In the case described above, Na$^+$ levels would return to their correct levels once the candidate tissue had been reimplanted in the animal concerned.

(iii) Administration of an extract of the tissue or organ should also result in reversal or prevention of the associated deficiency symptoms.

(iv)   Finally, the suspected hormone must be purified, its structure determined and tested for biological activity. It must exert the same biological effect as that seen previously with the intact organ or tissue.

4.4 The chemical nature of hormones

Virtually all hormones from both invertebrate and vertebrate animals, fall into one of three major classes – peptides or proteins, amino acid derivatives and steroids. There are exceptions to this, such as the range of C_{20} compounds known as the prostaglandins. The compounds serve many functions in animals and are beyond the scope of the present text. Representative examples for all three major classes and an example of a prostaglandin are shown in *Figure 4.3*. The chemical nature of the hormone is important because ultimately it decides how the hormone exerts its biological effect.

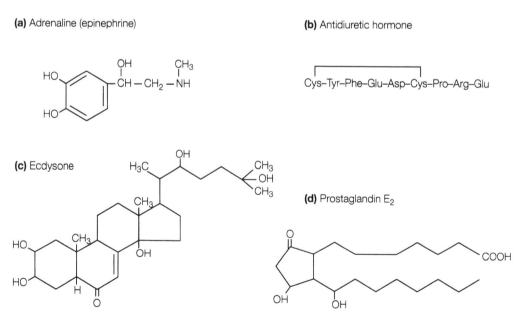

Figure 4.3. Examples of the major chemical classes of hormones. (a) An amino acid derivative; (b) a peptide hormone; (c) a steroid hormone; (d) a prostaglandin.

4.5 The mechanism of hormone action

For any hormone to exert its biological effect it must interact with its own specific **receptor**. The receptor, usually a large protein molecule, has a unique three-dimensional shape that will only bind a particular hormone or analogs of that hormone (compounds which possess a chemical structure which is very similar to the hormone concerned). As will be seen later, the receptor sites for hormones are located in either the cell membranes or the **cytoplasm** of cells. The specificity of hormone action, that is, the fact that only particular cells will be affected by a particular hormone, is determined by the presence or absence of receptors for that hormone in a given cell. If the receptor is absent from a cell then that cell is unresponsive to the hormone. This explains why hormones released into the general circulation have effects on specific cells, rather than more general effects on all cells as may be expected by their presence in the body fluids as a whole.

4.5.1 Membrane bound receptors

Hormones which have a peptide or protein structure and most of the amino acid derivatives combine with receptors that are located in the plasma membrane of the target cells. As will be seen later, the hormone–receptor interaction results in some change in cellular function. Such molecules – peptides, proteins and amino acid derivatives – are hydrophilic in nature, easily dissolving in aqueous solvents such as body fluids. This implies that their solubility in organic solvents is low. In order to exert their effect and produce a biological response, the hormone somehow has to influence the functioning of the target cell. Given the chemical nature of these hormones, they would be unable to cross the plasma membrane to influence cellular processes and therefore require a receptor located in the cell membrane.

The hormone binds to its membrane receptor, rather like a key fits into a lock. In doing so, it is said to produce an activated receptor which triggers a series of biochemical reactions that leads to the production of a biological response (*Figure 4.4*). The initial step in this series of reactions is the activation of another membrane bound protein called a **G protein**. (In actual fact there are many different types of G protein, but that need not be of any concern at the moment.) G proteins are actually trimers, i.e. they consist of three subunits. One of these subunits binds the substance **guanosine diphosphate (GDP)** at rest, hence the name G protein. However, activation of the G protein by a hormone–receptor complex causes the phosphorylation of GDP to produce **gaunosine triphosphate (GTP)**. This produces a **conformational change** in the G protein and, as consequence, the G protein dissociates, splitting into its constituent subunits. The subunit which has the GTP bound to it activates another membrane bound enzyme

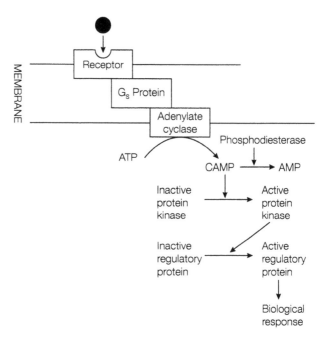

Figure 4.4. A summary of the biological events occurring when a hormone combines with a membrane-bound receptor. The ultimate response is the phosphorylation of a protein which in turn leads to some alteration in the biological activity of the cell concerned. G_s protein, stimulatory G protein.

called **adenylate cyclase**. GTP is eventually converted back to GDP by the GTPase activity of the G protein, and as a consequence the G protein returns to its original conformation. This process takes several seconds and allows the activation of many molecules of adenylate cyclase. Adenylate cyclase removes phosphate groups from ATP to form **cAMP** (cyclic adenosine monophosphate) (*Figure 4.4*). The function of cAMP is to activate another enzyme called a protein kinase. Having done this, the cAMP is broken down to AMP by the enzyme phosphodiesterase. Kinases are enzymes which phosphorylate other molecules, and the activated protein kinase 'looks' for a substance to phosphorylate. It is this final phosphorylation which produces the biological response to the initial hormone binding to the cell. The substances phosphorylated are proteins, and the effect of phosphorylation is to alter their conformation. If the protein is, say, an ion channel, then phosphorylation may well cause it to alter from its closed state to its open state. This will result in the movement of ions across the membrane which gives the desired response. For example, Ca^{2+} ions may enter the cell via Ca^{2+} channels and the resultant increase in intracellular calcium concentration may, for example, cause the cell to secrete some substance, or to contract – whatever response is appropriate for that cell. Listed below are other processes which may be altered by hormones in this way.

- Activation of enzymes, e.g. certain metabolic pathways may be switched on.

- Activation of **active transport** mechanisms, e.g. substances may be taken up into the cell.
- Activation of microtubule formation. This may be the initial step in the secretion of substances.
- DNA metabolism may be altered. This may be important in the growth or division of cells.

In the process described above, cAMP is termed a second messenger molecule, the hormone being the first messenger. It is now known that many other substances are able to function as second messenger molecules within cells. For example, hormone–receptor interaction may result in the activation of the membrane-bound enzyme phospholipase C. This enzyme in turn promotes the breakdown of the membrane lipid phosphatidyl inositol bisphosphate (PIP_2) into diacylglycerol (DAG) and inositol triphosphate (IP_3), both of which function as second messenger molecules. IP_3 diffuses into the interior of the cell where it promotes the release of Ca^{2+} from intracellular stores, e.g. the endoplasmic reticulum, or the sarcoplasmic reticulum in muscle cells. DAG remains in the membrane and in turn activates another enzyme called protein kinase C. Protein kinase C functions in the same way as the kinases described previously – once activated it will phosphorylate a substrate molecule. In this process, it is possible to consider the released Ca^{2+} as a third messenger molecule because it goes on to mediate several biological effects.

4.5.2 *Hormones using cytosolic receptors*

The steroid hormones and some amino acid-derived hormones (e.g. **thyroxine** released from the thyroid gland in mammals), utilize cytosolic receptors as opposed to membrane-bound receptors. These hormones are highly lipid soluble and pass very easily across the plasma membrane of target cells. There is some debate as to how these hormones produce a biological response, but it is thought that the hormone arrives at the target cell in conjunction with some sort of carrier molecule. Since the molecule is lipid soluble, it is unable to dissolve in the aqueous body fluids of animals. Thus, it combines with a carrier molecule, and it is this hormone carrier complex which is transported around in the body fluids of the animal. The hormone dissociates from the carrier molecule and freely enters the target cell. In the cytoplasm of the target cell the hormone combines with its specific receptor. Interaction between the hormone and its receptor results in an activated hormone–receptor complex. The activated receptor–hormone complex has a high affinity for **DNA**. This complex enters the nucleus where it combines with receptors associated with the DNA and it initiates changes in **DNA transcription**. The nature of the receptor site on the DNA molecule is not fully established, but it is thought that binding takes place in a region of the DNA known as the promoter region. By binding to this region of the DNA, it is possible to switch a particular gene on or off.

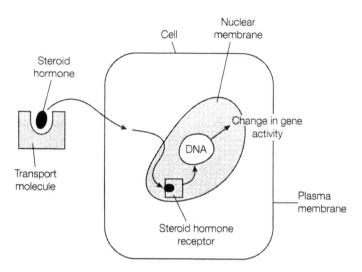

Figure 4.5. A summary of the signaling mechanism of steroid hormones. The hormone–receptor complex found in the nucleus binds to a specific region of DNA and alters the expression of particular genes. Some steroid hormone receptor proteins may be located in the cytoplasm.

Overall, the function of steroid hormones is to stimulate or repress the production of proteins – they are capable of turning genes on and off. The proteins that are then produced modify biochemical processes in the cell, thus producing the biological effect of the hormone. For example, the protein produced may be an enzyme that influences the metabolism of the cell. This may mean that a particular metabolic pathway within the cell can be switched on or off. Although, in some senses, it is a simpler way of producing the biological effect of the hormone, the action of steroid hormones tends to be slower than that of hormones which have a membrane-bound receptor. The mechanism of steroid hormone action is summarized in *Figure 4.5*.

4.6 Invertebrate endocrine systems

Invertebrate animals rely heavily upon neuroendocrine control systems as opposed to classical endocrine systems. However, as invertebrates have become more and more complex, both in structure and physiology, so have the physiological functions controlled by the endocrine system. This is partly because higher invertebrates, e.g. molluscs, have better-developed circulatory systems than lower invertebrates such as flatworms, and they therefore have a more efficient mechanism for distributing the necessary hormones. A brief summary of endocrine function in selected invertebrate phyla follows.

4.6.1 Coelenterates

Coelenterates, such as *Hydra*, have cells which secrete substances involved with reproduction, growth and regeneration. If the head of a *Hydra* is removed a peptide molecule is secreted by the rest of the body. This substance is called the 'head activator'. Its effect is to cause the remainder of the body to regenerate the mouth and tentacles which make up the head region.

4.6.2 Platyhelminthes

In a similar manner to the coelenterates, substances have been found in the flatworms which are involved in the processes of regeneration. It has also been suggested that hormones are involved in osmotic and ionic regulation as well as in reproductive processes.

4.6.3 Nematodes

Evidence for a role for neuroendocrine control in the nematodes has come from those members of the phyla which are parasitic. As with many parasitic organisms, different stages of the life cycle are quite often completed in different hosts. This means, therefore, that developmental changes in the nematode must coincide with the movement of the nematode to a new environment or new host. Some nematode worms may need up to four moults of their cuticle in order to complete their life cycle. It has been established that nematodes have neuroendocrine secretory structures associated with their nervous systems, in the ganglia in the head region and some of the nerve cords which run the entire length of the body. It is possible that changes in the immediate environment of the worm are the triggers for activity in this neuroendocrine control system and the consequent changes in physiology which are necessary for these animals.

4.6.4 Annelids

Annelid worms, such as polychaetes (e.g. *Neris*), oligochaetes (e.g. *Lumbricus*) and the hirudinae (e.g. leeches) show a reasonable degree of cephalization. The brain has been shown to contain large numbers of neurosecretory neurons, and these animals also have a reasonably well developed circulatory system. Therefore, the importance of endocrine control in these animals is enhanced. The endocrine systems in these animals are associated with such activities as growth and development, regeneration and development of the reproductive system. A good example of this is the metamorphic transformation of adult polychaete worms, known as **epitoky**, whereby some body segments become reproductive structures. Some of these transformed body segments break off altogether to become free-living organisms, as seen in some annelids. This process is known as **stolonization**. Epitoky is controlled by a neuroendocrine control system, but the hormone which is released actually inhibits this process. Therefore, during epitoky the levels of this hormone must be reduced, otherwise epitoky will be inhibited. How this works is unclear, but secretions may be regulated by environmental cues as some worms have breeding seasons. Similarly, there is a hormone that inhibits the development of gametes, and it is possible that it is the same hormone that controls epitoky.

4.6.5 Molluscs

It is known that there are large numbers of neuroendocrine cells in the ganglia which constitute the central nervous system of molluscs, particularly in the snails. It is also likely that there are some classical endocrine organs in addition to the neuroendocrine cells. Many of the substances released appear to be protein-like. They control many functions, such as osmotic and ionic regulatory processes, the regulation of growth and reproduction. Much of what is known about the molluscan endocrine system has come from studies on the reproductive systems of these animals. Reproduction in molluscs is complicated since many are hermaphroditic, i.e. are male and female simultaneously. A further complication is that some species display protandry, whereby the male gametes appear before the female gametes. Some of the substances utilized as hormones have been identified; for example, there is a hormone which stimulates the release of eggs from gonadal tissue and egg laying. In the cephalopods, where there are distinct males and females, reproduction is also under endocrine control. In this case, though, the role of classical endocrine glands, particularly structures called the optic glands, is thought to be important. The optic glands are thought to secrete several hormones which are required for sperm and egg development.

4.6.6 Crustaceans

Like the invertebrates described so far, the crustaceans have a predominantly neuroendocrine control system. However, they also have classical endocrine organs. The range of physiological functions controlled by the endocrine system is more varied and includes such aspects as osmotic and ionic regulation, heart rate regulation, blood composition, growth and moulting. Neuroendocrine control is best developed in the malacostrans (e.g. crabs, lobsters, shrimps). The neuroendocrine cells of these crustaceans are located in three major regions:

(i)   the sinus gland complex, sometimes called the X organ–sinus gland complex – this receives neuroendocrine axons from the ganglia of the head and optic lobes and is located in the eye stalks;
(ii) the postcommissural organ – again, this receives axons from the brain which terminate at the beginning of the esophagus;
(iii) the pericardial organ – this receives axons from the thoracic ganglia, and is located in close proximity to the heart.

In addition, there are a small number of classical endocrine organs. The Y organ is a pair of glands, located in the thoracic region of the animal on the maxillary or antennary segment. The secretions of the Y gland are thought to be involved in the process of molting – the periodic shedding of the exoskeleton as the animal grows. The mandibular gland located near to the Y gland is also thought to serve an endocrine function. There are also endocrine structures associated with the repro-ductive organs of these animals. For example, the androgenic gland is believed to be involved in the development of the testes and sperm production. The location of these glands is summarized in *Figure 4.6*.

An example of endocrine control in crustaceans is the ability of many crustaceans to change color. The advantage of this is that it enables them to assume the color of their background, thus helping them to avoid detection by predators, for example. The ability to change color varies from species to species; for example, some can only alter the shade of their coloration, e.g. from light to dark, whilst others can assume up to six different background colors. Color change is effected by the dispersal of pigments in cells called chromatophores. These cells are mainly located in the body covering, but they may also be found in some deeper organs. A variety of endocrine substances have been shown to alter chromatophore function. It is known, for example, that the sinus gland complex can release a pigment-concentrating hormone and a pigment-dispersing hormone, both of which are peptide hormones. It is also thought that some hormones are released from the pericardial organ which influence chromatophore function.

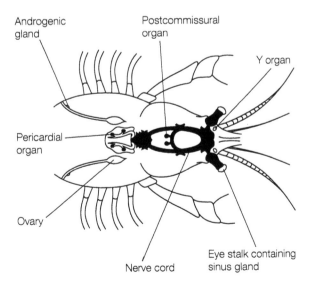

Figure 4.6. The major endocrine glands of a typical crustacean. Adapted from Withers, P., *Comparative Animal Physiology*, 1992, Saunders College Publishing.

4.6.7 Insects

Insects are similar to crustaceans in the wide variety of physiological functions that are controlled by endocrine organs (in comparison to other invertebrate phyla) and the predominance of neuroendocrine control systems. There are three main groups of neuroendocrine cells in the nervous system of insects:

(i)   the median neurosecretory cells, which send axons down to the paired corpora cardiaca which act as storage and release sites for the neurohormones;
(ii) the lateral group of neurosecretory cells, which also send their axons down to the corpora cardiaca;
(iii) the subesophageal neurosecretory cells, which send their axons down to the corpora allata, which are classical endocrine glands.

Insects also possess classical endocrine glands. The corpora allata are an example of classical endocrine organs, even though they are under neural control as described above. Another classical endocrine gland of importance is the prothoracic gland, which is located in the thorax of the more advanced insects and in the head region in less advanced insects. The locations of these structures are shown in *Figure 4.7*.

Growth is one of the physiological functions controlled by the endocrine system in insects. The growth of insects occurs in a stepwise fashion that requires a series of molts of the exoskeleton. In some insects, e.g. the

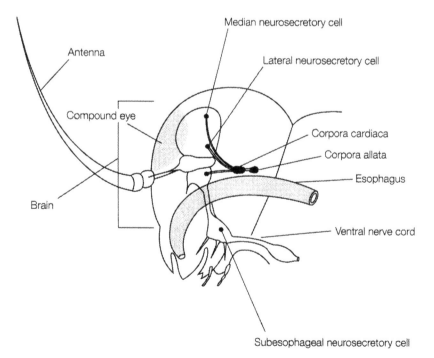

Figure 4.7. A simplified view of the insect nervous and endocrine systems.

collembola, molting continues even when the animal has reached maturity. In the majority of insects, however, there is a final molt when the juvenile is transformed into the adult. This process is known as **metamorphosis**. In some insects, e.g. the cockroach, the juvenile stages are similar in appearance to the adult animal. This type of development is called **hemimetabolous development**. In other insects, such as butterflies and moths, the juvenile stages are quite different from the adult animal, and this is known as **holometabolous development**. The process of molting involves several stages. First, the epidermal cells of the old cuticle withdraw and increase in number by the process of **mitotic cell division**. A new cuticle is produced whilst the old cuticle is digested and absorbed by the epidermal cells. The new insect then appears from the old cuticle, a process known as ecdysis, and, finally, the new cuticle hardens. This whole process is under endocrine control. The median neurosecretory cells produce a peptide molecule called prothoracicotropic hormone (PTTH, of which there are several types), which is released from the corpus cardiaca. This in turn stimulates the prothoracic gland to release the hormone ecdysone, which causes moulting. Another hormone, known as juvenile hormone, which is responsible for controlling metamorphosis is released from the corpus allata. High levels of this hormone result in the development of an immature animal, which indicates that there must be a close association between ecdysone and juvenile hormone if development is to proceed correctly – juvenile hormone secretion must be inhibited before the final molt in animals undergoing metamorphosis, although how this is achieved is not fully understood.

4.7 Vertebrate endocrine systems

In contrast to the invertebrate endocrine system, the emphasis in the vertebrate endocrine system is on classical endocrine organs with many physiological processes controlled by these organs. However, the nervous system still exerts an influence over the endocrine system since some of the peripheral endocrine organs are under the control of the anterior pituitary, which will be described later. During vertebrate evolution, there has been much conservation in terms of endocrine function. This means that some hormones have found new roles – for example, the hormone thyroxine controls **metabolic rate** in mammals, but in amphibians it is essential for the metamorphosis from tadpole to adult frog. In addition to this, as the vertebrates have evolved, new endocrine organs have emerged, such as the parathyroid glands that control Ca^{2+} levels which first appeared in the teleosts (bony fish). The typical vertebrate endocrine system is seen to consist of three principal glands or groups of glands:

- the hypothalamus;
- the pituitary gland;
- peripheral endocrine glands.

4.7.1 The hypothalamus and pituitary gland

The hypothalamus is part of the vertebrate brain and sits beneath the thalamus. Its main function is as an interface between the nervous and endocrine systems. A major role of the hypothalamus is to control the pituitary gland – the so-called master gland. The relationship between the hypothalamus and pituitary gland is shown in *Figure 4.8*. The secretions of the hypothalamus are transported to the pituitary gland. There are two types of secretions – those that are released into the posterior pituitary gland and those released into the anterior pituitary gland. Hormones secreted by the hypothalamus travel down axons extending from the hypothalamus to the posterior pituitary gland (the neurohypophysis). This region has a typical neuroendocrine role in that hormones are released from the posterior pituitary gland directly into

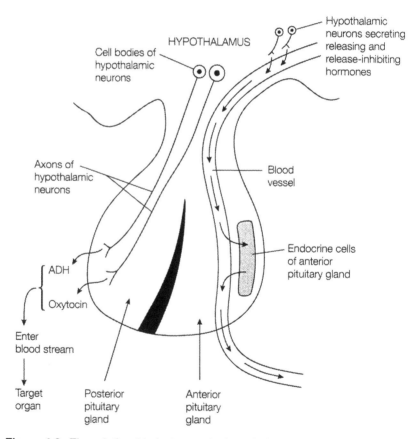

Figure 4.8. The relationship between the hypothalamus and the pituitary gland. Hormones are synthesized in the cell bodies of the hypothalamic neurons. ADH and oxytocin are released from the axon terminals into the posterior pituitary gland. The releasing and release-inhibiting hormones are released from axon terminals into capillaries and are transported to the anterior pituitary via the blood. These hormones stimulate the release of other hormones from the endocrine cells of the anterior pituitary.

the circulation. In mammals, the hormones released from the posterior pituitary gland are ADH (also known as vasopressin), which controls water absorption in the kidney, and oxytocin, which stimulates uterine smooth muscle contraction and milk ejection from the mammary glands. Both of these substances are peptides. Peptides with different amino acid compositions, but which have a similar biological role to either ADH or oxytocin, are found in all vertebrates. Another important group of secretions produced by the hypothalamus are the releasing hormones. These substances are released from axon terminals into **capillaries** which then pass to the anterior pituitary gland (adenohypophysis). Releasing hormones are thus delivered to the anterior pituitary gland indirectly via the blood system, rather than by direct release from axon terminals. The function of the releasing hormones, as the name suggests, is to influence the release of hormones from the anterior pituitary. The hormones released from the anterior pituitary then influence the secretions from other structures. Alternatively, release of hormones may be inhibited by release-inhibiting hormones secreted by the hypothalamus. The hypothalamic releasing and release-inhibiting hormones and the hormones whose release they control are shown in *Table 4.1*. It is essential that the plasma concentrations of all the secretions in this system are maintained at acceptable levels. The secretion levels are controlled via a negative feedback mechanism which is summarized in *Figure 4.9*.

Table 4.1. Hypothalamic hormones and the anterior pituitary hormones they influence

Hypothalamic hormone	Anterior pituitary hormone influenced
Releasing hormones	
Growth hormone-releasing hormone (GHRH)	Growth hormone
Thyrotropin-releasing hormone (TRH)	Thyrotropin-stimulating hormone (TSH)
Prolactin-releasing hormone (PRH)	Prolactin
Luteinizing hormone-releasing hormone (LHRH)	Leutenizing hormone
Follicle stimulating hormone-releasing hormone (FSHRH)[a]	Follicle-stimulating hormone
Melanocyte-stimulating hormone releasing hormone (MSHRH)	Melanocyte-stimulating hormone
Corticotropin-releasing hormone (CRH)	Corticotropin (adrenocorticotropic hormone, ACTH)
Release-inhibiting hormones	
Growth hormone release-inhibiting hormone (GHRIH, somatostatin)	Growth hormone
Prolactin release-inhibiting hormone (PRIH)	Prolactin
Melanocyte-stimulating hormone release-inhibiting hormone (MSHRIH)	Melanocyte-stimulating hormone

[a]FSHRH and LHRH may be identical substances and are sometimes referred to as gonadotropin releasing hormone (GnRH)

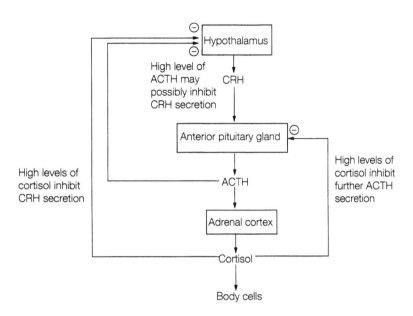

Figure 4.9. Diagram illustrating the control of cortisol secretion. Reduced levels of cortisol and ACTH have the opposite effects to those shown in the diagram above.

Table 4.2. Some established endocrine organs and the functions of the hormones they produce

Endocrine organ	Hormone released	Hormone function
Parathyroid gland[a]	Parathormone/calcitonin	Increase/decrease of blood Ca^{2+} levels
Stomach[b]	Gastrin	Regulation of acid secretion
Adrenal medulla[c]	Adrenaline	Short term response to stress, e.g. increases blood sugar levels, increases cardiac output
Adrenal cortex	Glucocorticoids, e.g. corticosterone	Regulation of metabolism
	Mineralocorticoids, e.g. aldosterone	Regulation of electrolyte levels
Ovary	Estrogens	Initiates proliferation of endometrium
	Progestogens	Supports thickened endometrium
Testis	Androgens, e.g. testosterone	Maintains production of gametes and is involved in the development of secondary sexual characteristics

[a]Except fish, there is no secretion of parathormone and a calcitonin-like substance called hypocalcin is secreted by the corpuscles of Stannius
[b]Gastrin is not produced in the lower vertebrates, e.g. fish and amphibians
[c]In higher vertebrates the adrenal medulla and cortex constitute a single gland, in lower vertebrates they are separate

Growth hormone promotes growth in all vertebrates. It has effects on carbohydrate, lipid and protein metabolism. It also induces the liver to release a compound called somatomedin which stimulates mitosis in bone tissue. Thyrotropin releasing hormone stimulates the thyroid gland to secrete thyroxine and triiodothyronine. The secretions of the thyroid gland have a variety of effects – the control of metabolic rate in mammals and the control of metamorphosis in amphibians. Prolactin is a hormone which is well known for its effects on reproductive tissue, and its stimulatory effect on milk production. However, it also has other effects, influencing water and Na^+ exchanges in amphibians. Follicle-stimulating and luteinizing hormones (FSH and LH) affect the gonads. FSH promotes gamete (i.e. egg and sperm) development, whilst LH, amongst many other functions, promotes steroid production. Melanocyte-stimulating hormone is involved in physiological color change in some of the lower vertebrates, e.g. amphibians and fish, whilst in some higher vertebrates it may be involved in osmotic and ionic regulatory processes. Adrenocorticotropic hormone stimulates the adrenal cortex to release the hormones produced there (i.e. the mineralocorticoids, such as cortisol).

4.7.2 Peripheral endocrine organs

Within the vertebrate phyla, there are a vast array of organs which have an established endocrine function. The list is ever increasing; it is now known that the heart produces a hormone (atrial naturetic peptide, ANP), which is involved in the renal regulation of Na^+. *Table 4.2* shows a few examples of some of the more established endocrine organs and their secretions in the vertebrates. It should not be considered a complete list and the reader is directed to more comprehensive endocrinology texts for further information.

Ventilation and gas exchange

5.1 Introduction

Ventilation, or gas exchange, is the exchange of O_2 and CO_2 between the body fluids of an animal and the environment in which that animal lives. The O_2 that an animal removes from its environment is used for the production of ATP by the oxidation of foodstuffs. Associated with this is the production and release of CO_2 back into the environment. This production of ATP is termed metabolic respiration. Virtually all animals depend on such aerobic metabolism to satisfy their resting energy requirements, although it is possible to produce energy, in the form of ATP, by anaerobic metabolism. The advantage of aerobic metabolism, compared with anaerobic metabolism, is that it significantly enhances the amount of ATP which can be made available. Consider the metabolism of a single molecule of glucose. During aerobic metabolism, 38 molecules of ATP are produced, whereas during anaerobic metabolism only two molecules are produced. Furthermore, it is possible to metabolize other substrates, such as lipids and proteins aerobically, but not anaerobically.

The exchange of gases between animals and their environments occurs by the process of simple diffusion. For many animals, small aquatic animals in particular, the exchange of gases across the general body surface is sufficient to meet the demands of the animal. However, as animals have evolved, exchange of gases across the body surface has become inadequate. Several evolutionary changes placed pressure on gas exchange across the general body surface area. For example, metabolic rates have tended to increase as animals have evolved, increasing the oxygen requirements of animals. The evolution of multicellular animals has resulted in specialization of cells located in the general body surface. The overall effect of these pressures has been to reduce the area that is available for gas exchange. Furthermore, the resultant increase in the complexity of animals has required a progressively more efficient gas exchange mechanism to be developed. The result of these pressures has been the development of specialized ventilatory structures.

5.2 Gases in air and water

By and large, animals tend to be either air breathers or water breathers. There are a few exceptions to this generalization; for example, larval amphibians (tadpoles) are water breathers, whilst adult amphibians are air breathers, and the common eel (*Anguilla*) is capable of breathing both air and water. Before going on to discuss the biology of ventilation, it is necessary to have an understanding of the physical and chemical principles relating to gases in air and water and the process of diffusion.

5.2.1 Gases in air

Normal, dry atmospheric air has the following composition: 20.95% O_2, 0.03% CO_2 and 78.09% N_2. The remainder of the volume is made up of the noble gases (e.g. argon, krypton, neon) which, together with N_2, are considered to be physiologically inert. Atmospheric pressure at sea level is 101 kPa (or 760 mmHg). At this pressure and at 0°C, one mole of gas occupies 22.4 l – this is the standard temperature and pressure (STP). Remember that one mole of a gas is the quantity that contains Avogadro's number (i.e. 6.02×10^{23}) of molecules. Gas volumes are usually expressed at STP, the reason being that it allows direct comparison of results obtained at different pressures and temperatures. In a mixture of gases, such as air, the total pressure exerted by the gas mixture is the sum of the pressures of the individual gases. This is known as Dalton's law of partial pressures. The **partial pressure** (*p*) of any gas is calculated by multiplying the percentage composition of that gas in the mixture by the total atmospheric pressure. Thus, the partial pressure of O_2 is calculated as follows: $pO_2 = 20.95/100 \times 101 = 21.1$ kPa. The same principle follows for CO_2 and N_2, giving pCO_2 as 0.03 kPa and pN_2 as 78.8 kPa. So far, only dry gases have been considered. In reality, atmospheric air always contains some water vapor. This water vapor also exerts a partial pressure, known as the water vapor pressure. The water vapor pressure increases with temperature. Thus, at 0°C, water vapor pressure is 0.6 kPa, and at 100°C it is equal to 101 kPa. The presence of water vapor reduces the partial pressures of gases in the air. Thus, at STP, $pO_2 = 20.95/100 \times (101 - 0.6)$. At 37°C, the water vapor pressure is about 6 kPa.

5.2.2 Gases in water

Gases are soluble in water. When a sample of air and water are allowed to equilibrate with each other, the partial pressure of the gases in water will be the same as those in air. The solubilities of different gases in water vary considerably. For example, CO_2 is approximately 30-times more soluble in water than is O_2. It has been said previously that the partial pressure of gases in water is the same as that in the gas phase when the system is at equilibrium. However, the concentration of the gases in the water will not be the same. The concentration of gas in water is described by Henry's law, which states that:

$$[G] = pG \times \alpha_G$$

where [G] is the concentration of the gas in water (mol l^{-1}), pG is the partial pressure of the gas (kPa) and α_G is the solubility coefficient (mol l^{-1} kPa^{-1}). Thus, for the same partial pressure, the amounts of O_2 and CO_2 dissolved in water will not be the same because the solubility of the two gases is different. Other factors influencing the gas content of water include temperature and the presence of other solutes. Increasing temperature decreases the amount of gases dissolved in water; for example, there is a 40% reduction in the O_2 content of water as temperature increases from 10°C to 30°C. In a similar manner, the presence of solutes in water also decreases the gas content. At the same temperature, there is a 20–30% reduction in the amount of O_2 dissolved in seawater when compared to freshwater.

5.2.3 Diffusion

The movement of gases, across the general body surface or from alveoli to pulmonary capillary blood in the human lung occurs by the process of diffusion. That is, molecules of gas move from a region of high partial pressure to one of lower partial pressure down a concentration gradient. It is possible to attempt to quantify the process of diffusion by the use of Fick's law. In its simplest form, Fick's law defines the movement of a gas from one region to another in a single dimension; for example, a molecule of O_2 passing from water across the body surface into an animal. Fick's law is stated as:

$$\frac{J}{A} = \frac{-D(c_1 - c_2)}{x}$$

where J is the mass of gas transferred per unit time, A is the surface area available for gas exchange, J/A is the amount of gas moving per unit area of exchange per unit time, $c_1 - c_2$ is the concentration gradient, x is the distance over which diffusion occurs and D is the diffusion constant. The negative sign indicates that movement is occurring down the concentration gradient. D varies depending upon the gas, the nature of the solvent in which it is dissolved and the temperature of the solvent. The equation gives information regarding the physiological characteristics of diffusion – first, that the amount of gas diffusing is proportional to time, and second, if the diffusion distance is increased, the amount of gas that actually diffuses decreases.

5.3 A comparison of air and water as respiratory media

As stated previously, animals generally breathe air or water. What are the advantages and disadvantages of utilizing either of these substances? First, water is much more dense and viscous than air, some 1000-times more dense and 60-times more viscous. This means that water is, in metabolic terms, much more expensive to pass through or over a respiratory structure. Such is the metabolic cost of breathing, that a fish may expend up to 10% of its resting metabolism simply passing water over its gills. Obviously, when the fish becomes more active, the cost of ventilation increases even more. However, to be offset against this is the fact that air-breathing animals must expend energy to overcome the effects of gravity (aquatic animals do not have this problem). In order to limit the metabolic cost of ventilating gill structures, animals that breathe water usually have a unidirectional flow of water through the gills.

The oxygen content of water is much reduced when compared to air – water contains 10 ml of O_2 per liter, whereas air contains approximately 200 ml per liter. However, water has a CO_2 solubility 20–30 times greater than that of air. Thus, aquatic animals have very little problem in excreting the CO_2 produced during aerobic metabolism. The significance of this is that for aquatic animals O_2 is the primary stimulus for breathing, whilst in air breathers it is CO_2, simply because these are the 'rate-limiting' gases. The fact that in air-breathing animals CO_2 levels in the body are much higher has implications for acid–base balance, since CO_2 is an acidic gas.

Water has a higher thermal capacity than air (i.e. water is an effective heat sink), and the consequence of this is that aquatic animals have tremendous problems when it comes to thermoregulation, as any heat that is produced is rapidly dissipated to the water. It is difficult for such animals to do anything other than conform to the temperature of the surrounding water. Conversely, air has a poor thermal capacity, so it is possible for air-breathing animals to utilize the heat of respiration (i.e. the heat which is produced as a by-product of biochemical reactions) for the process of thermoregulation. This tends to give air the advantage over water as a respiratory medium. Certainly, air-breathing animals tend to have higher metabolic rates than aquatic animals, and this presumably confers the animal with some advantage over those with lower metabolic rates, although, in turn, this also produces potential problems – for example, the continual requirement for large amounts of oxygen, and the possible problems of overheating.

5.4 Gas exchange by simple diffusion across the general body surface area

Gas exchange occurring across the general body surface area is the means by which single cell animals (e.g. protozoa) and relatively simple multicellular animals achieve respiration. In virtually all cases, we are considering aquatic animals; however, such mechanisms also operate in higher animals, as will be seen later. A problem with such a mechanism is that as O_2 is taken into the animal, so a layer of oxygen-depleted water called a boundary layer surrounds the animal. This boundary layer is nonmoving and, together with the cells of the body surface, presents the major obstacle to the diffusion of gases. This means, in reality, that actively-metabolizing cells must be within 1 mm of the aqueous environment. Multicellular animals have consequently had to change their morphology in order to survive without the development of specialized respiratory structures. Thus, jellyfish, which may grow in excess of 1 m in diameter, are arranged such that actively-metabolizing cells are placed around the periphery of the animal in direct contact with the water. Flatworms, which also lack any specialized respiratory structure, have become thin and elongated.

Gas exchange across the general body surface area is characteristic of the Porifera, coelenterates, the majority of Platyhelminthes, nematodes, rotifers and some annelids. This method of gas exchange is also known to occur in many vertebrates. For example, virtually all amphibians depend, to a greater or lesser extent, on gas exchange across the body surface. Some amphibians (e.g. the hairy frog *Astylosternus robustus*) develop vascularized hairs on the skin which increase the surface area available for gas exchange. The ability to exchange gases across the body surface is also well developed in some fish (e.g. eels). Generally, this method of gas exchange is less important in birds and mammals. These animals have high metabolic rates and, therefore, have high demands for oxygen. Gas exchange across the general body surface area would be inadequate to meet these demands. That said, there are a few examples where the general body surface area does make a contribution to the process of gas exchange. Bats, for example, may excrete between 10 and 15% of their CO_2 across the skin of the wing.

5.5 The evolution and design considerations of gas exchange organs

It is apparent that gas exchange across the body surface will only meet the requirements of small animals or those with very low metabolic rates. Some of the evolutionary pressures on the development of gas exchange organs were discussed in Section 5.1. Before the organs themselves are discussed, it is necessary to appreciate the functional requirements of such organs.

(i) They require a large surface area. This simply increases the amount of gas that can be moved from the environment to the animal and vice versa.

(ii) The separation between the blood of the animal and the air or water it breathes must be minimal. In many situations, body fluids and respiratory media are separated by the thickness of a single plasma membrane. This ensures a path of least resistance in terms of diffusion of the gases across it. However, the necessity for the organ to be thin must be tempered against a requirement for it to be structurally sound.

(iii) The movement of the respiratory medium must match that of the body fluids (into or out of which the gases move) to give the appropriate **ventilation:perfusion ratio**. As an example, human lungs each have what can be considered as their own blood supply. For the lungs to function effectively, they need to be both ventilated and perfused and it is therefore necessary to have the correct ventilation:perfusion ratio. A lung that is ventilated, but receives no blood supply will not be an effective gas exchange organ. In this case, there is an extreme mismatch of the ventilation:perfusion ratio. Ventilation:perfusion ratios only apply to lungs and gills. As we shall see later, the development of the tracheal system in insects has resulted in a gas exchange system which is independent of the circulatory system, so the above consideration does not apply.

(iv) The final aspect of the design of gas exchange organs is the physical arrangement between the respiratory medium and the body fluids. Several options are possible and these are illustrated in *Figure 5.1*.

- Uniform pool arrangement, the prime example of which is the mammalian respiratory system. In this situation, the respiratory medium (i.e. air) does not flow in any particular direction in relation to the flow of blood through the lungs. An equilibrium is established between gas concentrations in air and blood.
- Countercurrent arrangement, as seen in fish gills. Here, the flow of the respiratory medium (i.e. water) and blood are in

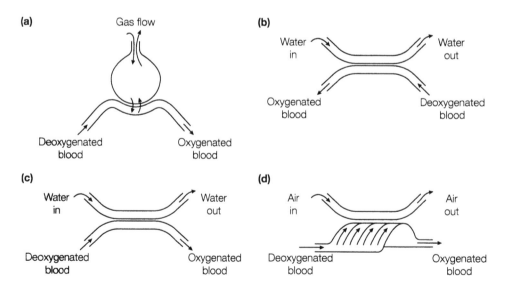

Figure 5.1. Different arrangements of blood vessels and respiratory media. (a) Uniform pool arrangement, e.g. the mammalian lung; (b) countercurrent arrangement, e.g. fish gills; (c) concurrent arrangement; (d) crosscurrent arrangement, e.g. bird lungs.

opposite directions. This is a particularly good arrangement, in that the entire length of the gas exchange surface is utilized. At any point along the gill, gas concentration in the water is greater than that in the blood, so there is continual movement of gas from water to blood.

- Concurrent arrangement where the respiratory media and blood flow in the same direction. Such a system is less efficient. Maximal theoretical gas concentrations in the blood are never achieved because gas concentrations equilibrate before blood has had a chance to perfuse the entire length of the respiratory organ.
- Crosscurrent flow. This is the arrangement found in bird lungs. Here, blood and respiratory medium flow almost at right angles to each other. Once again, this ensures maximal transfer of gases between blood and the respiratory medium.

Bearing in mind all the requirement of gas exchange organs, evolution has provided animals with three possibilities: gills, lungs and tracheal systems. Each of these will now be discussed.

5.6 Gills

Gills are the gas exchange organs of aquatic animals. They are outgrowths of the body surface which are highly folded or convoluted in order to maximize the surface area available for gas exchange. They may be secondarily enclosed in a protective structure, such as the opercular cavity in bony fish. In order for them to be ventilated, water must pass over their surface. This can be achieved in one of two ways: either the gill passes through water or water passes over the gill. The latter is only feasible for small organisms (e.g. some aquatic insect larvae), because of the excessive energy requirements that are required. Gills are present in both invertebrates and vertebrates.

5.6.1 Invertebrate gills

Gills are present in marine polychaete worms (e.g. *Arenicola*) (*Figure 5.2*). In these animals, the gills are modifications of the parapodia, which are lateral appendages. In the case of *Arenicola*, water is moved over the gills as a consequence of the general body movements of the animal. In other polychaetes, cilia may be present on the gill to ensure the flow of water. There is considerable diversity in the morphology of gills in the polychaetes which suggests that, in this group at least, gills have evolved several times. Molluscs also have gills and, although there is much variation in the gill, in all cases they are ciliated and this is the mechanism by which water passes over them. In some molluscs (e.g. the lamellibranch *Mytilus*), the gills have a secondary function in that they are involved in filter feeding.

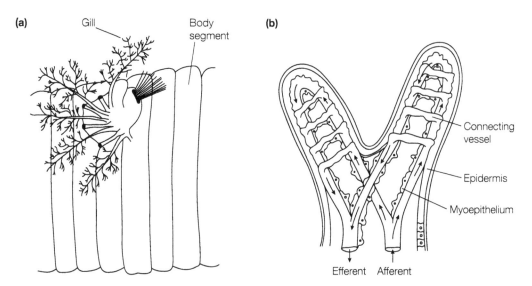

(a) Gill   Body segment   **(b)**

Connecting vessel

Epidermis

Myoepithelium

Efferent   Afferent

Figure 5.2. (a) The body wall of *Arenicola* sp. Showing the respiratory structures. Branching of the gills maximizes the surface area available for gas exchange. (b) An enlarged view of a gill. The arrows show the direction of blood flow. Redrawn from Jouin, C. and Toulmond, *Acta Zoologica* **70**, 121–129, 1989, with permission from The Royal Swedish Academy of Sciences.

In the case of crustaceans, it is mainly the malacostraca (crabs, lobsters etc.) which have gills. In these animals, the gills are usually derived from the thoracic or, occasionally, abdominal appendages and are usually enclosed within the carapace. The nature and number of gills is related to the environment in which the animal lives; for example, aquatic crabs tend to have larger gill areas than land crabs. Although water flow over the gills is usually unidirectional, many crustaceans have the ability to reverse this flow. The significance of this maneuver is that it rids the gills of accumulated debris.

As will be seen later, insects have developed a tracheal system to ensure adequate gas exchange, but many aquatic insects and aquatic larvae of terrestrial insects rely upon gills. Once again there are many variations on insect gills. In some cases they have developed from abdominal appendages, such as those found in Ephemeroptera larvae. Other variants include the development of gills within the lumen of the rectum, so-called anal gills. In such a situation, the gills are ventilated by the movement of water in and out of the rectum. This arrangement is seen in some dragonfly larvae.

5.6.2 Vertebrate gills

Many aquatic vertebrates depend, to differing degrees, upon gas exchange across the general body surface area. However, such a mechanism is virtually always insufficient to meet the needs of the animal concerned. Gills in vertebrates may be of two types, external

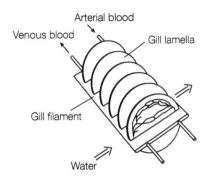

Figure 5.3. The structure of the gill in fish.

filamentous gills and internal lamellar gills, the latter being most common. Many vertebrates only have external gills during certain stages of development (e.g. fish and amphibian larvae). Internal gills are much more common. Gills are best developed and understood in teleost fish (*Figure 5.3*). They consist of several gill arches from which extend two rows of gill filaments. Upon each filament are rows of gill lamellae, which is the site of gas exchange. Water and blood flow in opposite directions, i.e. there is a countercurrent arrangement which maximizes O_2 uptake into blood. Water is moved over the gills by the pumping action of the mouth and the opercular cavity (*Figure 5.4*). Water is drawn into the mouth when the floor of the mouth is lowered (*Figure 5.4a*). The mouth then closes and water is forced over the gills, aided by the opening of the opercular flap which results in water being expelled (*Figure 5.4b and c*). This serves to maintain an almost continuous flow of water over the gills (*Figure 5.4d*). Many active, fast swimming fish, such as tuna and some sharks, swim with their mouth open and consequently force water over their gills. This is called ram ventilation. However, such a system does increase the energy requirement for swimming since it increases the drag of the fish in the water (i.e. it increases its resistance to flow through the water). Since ram ventilation is seen in many fish, particularly when swimming fast, it suggests that the increased energy requirements of swimming are less than the energy requirements of mouth and opercular pumping in order to ensure the delivery of water and oxygen to the gills. In teleost fish, gill size is related to activity, so that a fast swimming fish, such as mackerel, has a gill area ten-times that of the bottom-dwelling and much slower flounder.

5.7 Lungs

Lungs are ingrowths of the body surface, as opposed to gills which are outgrowths, and are the gas exchange organs of air-breathing animals. In a similar manner to gills, they are highly folded to increase the surface area available for gas exchange. They are intimately linked with

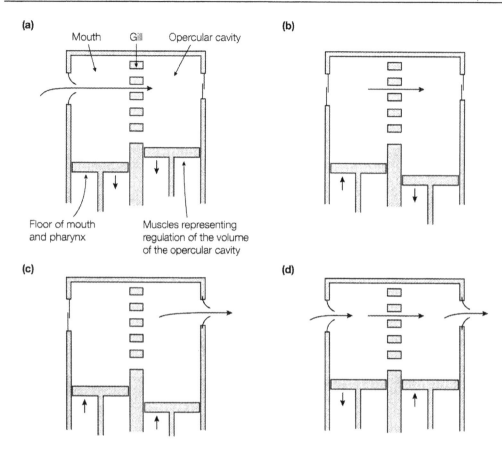

Figure 5.4. The flow of water through fish gills. Water movement is a function of pressure in the mouth and opercular cavity. An increase in the volume of the mouth of opercular cavity results in a reduction in pressure in these areas. Water will flow from a region of high pressure to a region of lower pressure down its pressure gradient.

the circulatory system in order to distribute gases around the animal. As will be seen later, lungs have evolved which work in various ways; for example, some lungs are actively ventilated, whilst others simply rely on diffusion to renew their gas content. In lungs that are actively ventilated, air may either be pumped or sucked into them. A major problem associated with breathing air is water loss and various solutions to this problem have evolved. However, it should be remembered that as water evaporates it cools, so water loss from the gas exchange organ may be important in terms of thermoregulation. This is particularly well developed in animals which do not possess sweat glands, e.g. dogs and birds.

5.7.1 Invertebrate lungs

Several invertebrate phyla have made the transition to a terrestrial way of life, some more successfully than others. Lungs have evolved in

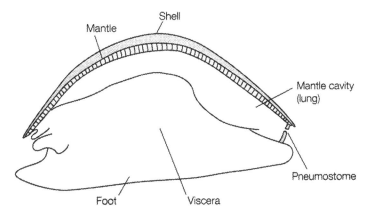

Figure 5.5. Generalized diagram of a mollusc showing the location of the lungs.

some classes of mollusc, e.g. the pulmonate snails. The lung is found at the back of such animals where the mantle cavity fuses with the rear of the animal (*Figure 5.5*) and it can be closed off to the atmosphere by closure of the pneumostome. It is capable of being ventilated, but exchange of gases between the environment and the lung generally occurs by simple diffusion. Crustaceans show varying degrees of success in their adaptation to terrestrial life and air breathing. Some amphipods which have invaded land retain gills, and their ability to live on land is due to behavioral adaptations (living in moist habitats, for example). Even those crustaceans which have evolved lungs or lung-like structures (e.g. woodlice), are still restricted to moist habitats. Other invertebrates, such as spiders and scorpions (i.e. chelicerates), also have lungs. In this case they are called book lungs, and they are located on the abdominal surface and consist of layers of lamellae, rather like those of fish gills (*Figure 5.6*).

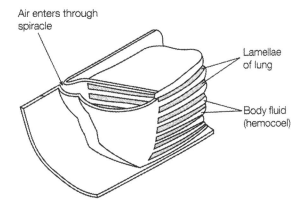

Figure 5.6. The structure of book lungs found in spiders. Air enters the spaces between the lamellae and gases pass into the body fluids.

5.7.2 *Vertebrate lungs*

Lungs are found in amphibians, reptiles, birds and mammals. Amphibians show tremendous diversity in the structure of their lungs, ranging from a simple noncompartmentalized cavity to well vascularized, highly compartmentalized lungs. However, when it comes to ventilating the lung they are all ventilated by positive pressure pumping. Air enters and leaves the mouth almost continually, as can be seen by the constant motion of the mouth in these animals. During this stage, air is prevented from entering the lungs by closure of the glottis. When lung ventilation occurs, the already open nostrils close, the glottis opens and air is forced from the mouth into the lungs under positive pressure. In many cases, this process occurs several times in order to fill the lungs with air. Removal of air from the lungs is achieved by opening the glottis, coupled with the natural elasticity of the lungs which return to their original volume, expelling air from the lungs.

The lungs of reptiles are morphologically more complicated than those of amphibians in that they are much more compartmentalized. A second major difference in reptilian lungs is that they are ventilated by a suction pump. Movement of the ribs outwards results in the development of a subatmospheric pressure in the thoracic cavity where the lungs are located. Air in the environment is at greater pressure than that in the lungs, so air moves inwards down its pressure gradient. Turtles and tortoises have enhanced this process even further by the development of a true diaphragm, which separates the thoracic and abdominal cavities.

Mammalian lungs are, in turn, more compartmentalized and morphologically more complex than reptilian lungs. Air is drawn into the lungs, again, by a suction pump mechanism. Air enters via the trachea, which, via a series of bifurcational divisions, terminates at the **alveoli**, the sites of gas exchange.

Finally, the respiratory system of birds should be mentioned. The anatomical organization of a bird's respiratory system is relatively complex. The lungs are connected to an extensive series of air sacs, which are known to extend into bone. These can be conveniently divided into two main groups: the posterior air sacs and the anterior air sacs. The arrangement of the lungs and air sacs is shown in *Figure 5.7*. Gas exchange in the bird lung takes place in the **parabronchi**, which are, essentially, a series of parallel tubes open at both ends and through which there is a unidirectional flow of air. The flow of air through the respiratory system of a bird is shown in *Figure 5.8*. It can be seen that two complete breathing cycles are required in order to move a single volume of gas through the system. The air sacs act as bellows, forcing air to move around the system.

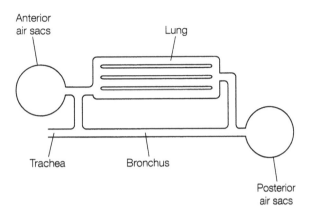

Figure 5.7. The generalized organization of the bird lung. Both the anterior and posterior air sacs represent collections of several, smaller air sacs. The lung consists of several flow-through tubes called parabronchi where the process of gas exchange occurs. Redrawn from Schmidt-Nielsen, K. *How Animals Work*, 1972, with permission from Cambridge University Press.

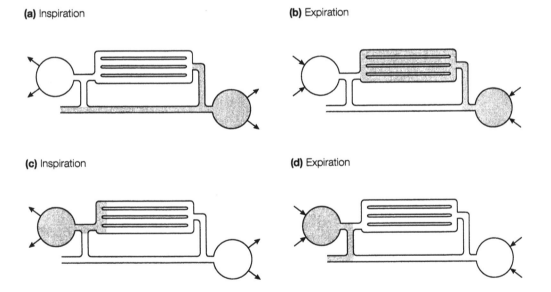

Figure 5.8. The movement of a single bolus of gas through the respiratory system of a bird. Arrows indicate movement of the air sac. The presence of anatomical valves ensuring one-way flow of air through the system has not been demonstrated. From Bretz and Schmidt-Nielsen, *Journal of Experimental Biology* 56, 57–65, 1972.

5.8 Tracheal systems

Gas exchange in insects occurs through a series of tubes called **trachea**, which branch dichotomously to give increasingly finer tubes called

tracheoles. The trachea connect to the environment through openings in the insect cuticle called **spiracles** (*Figure 5.9*). In principle, air is piped directly to the cells of the insect. In some cases, tracheoles are known to penetrate individual cells and terminate near mitochondria (*Figure 5.9(b)*). This arrangement makes the respiratory system independent of the circulatory system. The basic pattern is that twelve pairs of spiracles exist, three pairs of which are located on the thorax, with the remainder located on the abdomen. Generally, all the trachea interconnect with each other. This means it is possible to have a unidirectional flow of air throughout the animal. However, this basic pattern has been modified in many ways. For example, some insects, particularly large active insects (e.g. locusts), have developed air sacs. Air sacs are simply thin walled extensions of trachea. Since they are thin-walled, they are compressible, which means that their volume can be increased or decreased, so air can be pumped into and out of the system. In many aquatic insect larvae, the majority of the spiracles have become

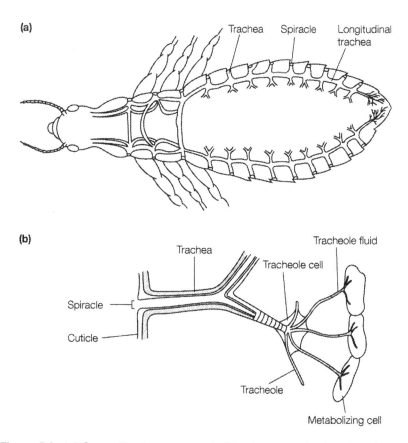

Figure 5.9. (a) Generalized arrangement of the insect tracheal system. Large trachea are interconnected. (b) The trachea divides into finer tubes called tracheoles which may enter individual cells. As demand for gas exchange increases, tracheole fluid is withdrawn to minimize the barrier to diffusion.

nonfunctional, with the exception of the last pair on the abdomen. In such cases, the larvae are suspended in water with the tip of the abdomen penetrating the surface, thus allowing gas exchange to occur. In the case of small, inactive insects, the air in the tracheal system enters and leaves by simple diffusion. In large, active insects, this is insufficient to meet the demands of the animal, hence the development of air sacs.

As with all air-breathing animals, loss of water can be a major problem. Insects have overcome this in a novel way known as **discontinuous** or **cyclic respiration**. With this, O_2 is taken up at a more-or-less continuous rate, but CO_2 is released in periodic bursts. In between bursts, CO_2 remains dissolved in body fluids or is buffered, and the spiracles are seen to flutter and then close completely. During fluttering, the pressure within the tracheal system becomes subatmospheric, thus ensuring the movement of air into the trachea and preventing air from leaving the insect. After fluttering, the spiracles close completely and O_2 from the air within the closed tracheal system is consumed. The significance of this arrangement is that it allows gas exchange and metabolism to proceed continually, whilst limiting the potentially dangerous loss of water from spiracles that would otherwise be permanently left open.

5.9 Control of ventilation

Breathing must be controlled in order to ensure that adequate gas concentrations exist in body fluids. Much remains to be discovered about the control of breathing in both water- and air-breathers. In general, aquatic organisms are more responsive to changes in O_2 whilst terrestrial organisms are more responsive to changes in CO_2. This arises because O_2 and CO_2 are the 'rate limiting' gases in aquatic and terrestrial environments, respectively. In the aquatic environment, CO_2 solubility is high, so its disposal presents no problems to animals. However, the O_2 content of water is relatively low. Therefore, the gas exchange system of an aquatic animal has developed to be significantly more responsive to changes in O_2 as opposed to changes in CO_2. In terrestrial animals, the opposite is true. The O_2 content of atmospheric air is relatively high, whilst the content of CO_2 is very low. Thus, the gas exchange system of a terrestrial animal is regulated by changes in CO_2.

As with any control system, there needs to be a means of measuring the required parameter (i.e. the gas concentration), a means of comparing actual gas levels to desired levels and, finally, some means of correcting any discrepancy. Even those aquatic animals that pass water over gas exchange organs with the aid of cilia may be able to modify ciliary action to ensure adequate gas exchange occurs. Animals that breathe with the involvement of somatic muscles have a more complicated control system, as described previously. For example, fish have a respiratory centre located in the medulla of the brainstem. Activity of these neurons produces the basic ventilatory pattern. Fish are able to monitor the level of O_2, although there is some dispute over the actual location of the receptors (in this case referred to as **chemoreceptors**). They may be present in the mouth, gills, opercular cavity or in the arterial or venous circulation. There are many other stimuli which affect the ventilatory pattern, for example, temperature, and osmotic changes. The regulation of ventilation in air-breathing animals follows the same principles as in aquatic animals, except that now CO_2 is the primary stimulus to breathing. Control is displayed even in the invertebrates. For example, insects have a sophisticated control system whereby CO_2 controls spiricular opening. There is also evidence that there may be a respiratory rhythm generated by neurons in the ventral nerve cord of some insects. In vertebrates, there is a region in the medulla of the brainstem where ventilatory rhythm is generated. Amphibians have chemoreceptors that respond to both O_2 and CO_2; however, the main stimulus is from CO_2. The precise control of breathing can vary depending upon the stage of the life cycle. In a similar manner, reptiles have CO_2 sensors in their respiratory system, although in some reptiles O_2, rather than CO_2, is the main stimulus for breathing. Like amphibians, birds also have a respiratory system which is primarily driven by changes in CO_2, although there are detectors which respond to changes in O_2.

The control of breathing is best understood in mammals. Once again CO_2, or, more precisely, its concentration in blood, is the main stimulus to breathe. Located on the surface of the medulla in the brainstem of mammals are chemoreceptors which are sensitive to CO_2. These receptors are bathed in cerebrospinal fluid, which is produced in the ventricles of the brain and flows over the entire surface of the brain and spinal cord. The role of the cerebrospinal fluid is to act as a 'shock absorber', providing the central nervous system with some mechanical protection. In reality, the chemoreceptors in the medulla are responsive to H^+ rather than CO_2 *per se*. This is because CO_2 in the blood passes into the cerebrospinal fluid and dissolves according to the following equation:

$$CO_2 + H_2O \leftrightarrow H_2CO_3 \leftrightarrow H^+ + HCO_3^-$$

This means that the central chemoreceptors are acting as pH meters. An increase in the CO_2 content of blood causes a decrease in the pH of cerebrospinal fluid (an increase in the H^+ concentration of a solution makes it more acidic). The response to this increase in blood CO_2 is an increase in the rate of breathing, thus exhaling the excess CO_2 and returning the blood CO_2 level to its normal value.

There are other receptors, located in large arteries, such as the aorta and carotid, which monitor blood O_2 levels. However, given that the primary stimulus to breathe is CO_2, the sensitivity of the chemoreceptors for O_2 is much less than the sensitivity of those receptors in the medulla which monitor CO_2. Therefore, the O_2 content of the blood must be reduced significantly before an increase in the rate of breathing is triggered in order to restore CO_2 concentration to its normal level.

Superimposed on this basic control are inputs from many other receptors located in the lungs and airways, the joints and elsewhere, together with influences from other brain regions. For example, receptors located in the joints may be important in signaling an increase in movement to the brain, such as occurs during exercise, for example. Obviously, during exercise there is a need for increased amounts of O_2 and the need to expel larger amounts of CO_2. Thus, increased activity of the joints is the stimulus for the breathing rate to increase.

Further reading

Jones, D. J. (1975) *Comparative Physiology of Respiration.* Edward Arnold, London.

Widdicombe, J. and Davies, A. (1983) *Respiratory Physiology.* Edward Arnold, London.

Thermoregulation in animals

6.1 The importance of temperature to animal physiology

Animal life exists at body temperatures which range between –2°C, in fish and invertebrates living in arctic waters, to +50°C in desert-living animals. Some creatures may exist at even more extreme temperatures; for example, it is now known that some polychaete worms live in deep sea vents at temperatures exceeding 80°C. For the vast majority of animals, though, the range of temperatures over which any individual can live is normally much narrower than this. Most animals simply assume the temperature of their immediate external environment, but birds and mammals regulate their body temperature and maintain it at a relatively constant level which is different to that of their immediate external environment. Temperature is important to animals for the reason that, within limits, an increase in temperature increases the rate of physical and chemical (i.e. metabolic) reactions. Temperature affects the kinetic energy of molecules, and as temperature and kinetic energy increase, the probability of individual reactant molecules colliding, and thus permitting the necessary reactions to occur, also increases. It is possible to quantify the increase in reaction rates by measuring the Q_{10} value. The Q_{10} is defined as the increase in the rate of a reaction or a physiological process for a 10°C rise in temperature. It is calculated as the ratio between rate of reaction (k) occurring at $(X + 10)$°C and the rate of reaction at X°C:

$$Q_{10} = \frac{k\ (X + 10°C)}{k\ (X°C)}$$

For most biological reactions, Q_{10} is between 2 and 3, i.e. rates of reaction double or treble. In contrast, the Q_{10} of physical processes, such as diffusion, is 1. Hence, it is possible to determine from the Q_{10} value whether a particular process is biological or physical. As the absolute temperature at which particular reactions occur increases, the value of Q_{10} will decrease. For example, the Q_{10} of a biological reaction measured between 20 and 30°C will be quite different from the Q_{10} for the same reaction measured between 70 and 80°C. This is because at higher temperatures the enzymes involved in the reaction will begin to denature, and this offsets the potential benefit of higher temperature in increasing the rate of reaction.

6.2 Classification of temperature regulation

As stated previously, the majority of animals assume the temperature of their external environment. Such animals are said to be thermoconformers. This contrasts with birds and mammals who maintain a body temperature different to that of their external environment. Such animals are said to be thermoregulators. On this basis, it is possible to classify animals according to the way by which they control their body temperature.

The simplest way to classify animals in the past has been as either cold-blooded or warm-blooded. This system classifies mammals and birds as warm-blooded and all other animals as cold-blooded and has little value in physiology. For example, a desert-living lizard (a cold-blooded animal) may well have a body temperature (at certain times) which is higher than a corresponding desert-living mammal (a warm-blooded animal). Also, some mammals that hibernate may well have very low body temperatures when compared with nonhibernating mammalian species. Alternative terms used for cold- and warm-blooded are poikilothermic and homeothermic, respectively. Poikilothermic means having a variable body temperature, whilst homeothermic means having a relatively constant body temperature. For the reasons given previously, the use of such terms provides little information about the temperature relationships of an animal. Perhaps the best way of classifying animals, therefore, is by dividing them into ectotherms and endotherms. An **ectotherm** is an animal which is dependent upon external sources for heat gain, whereas an **endotherm** is dependent upon internal heat production. Essentially, birds and mammals are endotherms and all other animals are ectotherms. This is a generally accepted classification, but even this is not without its anomalies. For example, insects are classified as ectotherms, but some insects e.g. those that fly, may generate additional heat by muscular contractions and are, therefore, partly endothermic.

6.3 Heat exchange interactions between animals and the environment

It is inevitable that animals will interact with their immediate thermal environment – there will be some heat exchange between the two. Although such interaction is inevitable, animals are able to manipulate this exchange to their advantage by using it as a way of regulating their body temperature in increasing or decreasing heat loss and gain. There are four possible ways by which animals can exchange heat with the environment: conduction, convection, radiation and evaporation.

A summary of the heat exchanges between an animal and its environment is shown in *Figure 6.1*. It is important to realize that maintenance of a desired body temperature, for both endotherms and ectotherms, is matter of balancing heat gain and heat loss.

6.3.1 Conduction

Conduction of heat is the transfer or movement of heat between two bodies that are in direct physical contact with each other. Mathematically, conductive heat transfer can be described in the following equation:

$$Q = kA\,(T_2 - T_1)/d$$

where:
Q = conductive heat loss,
k = thermal conductivity of the material through which heat is transferred, a measure of the ease with which heat may be transferred,
A = the area over which heat is transferred, and
$(T_2 - T_1)/d$ = temperature gradient between the two bodies (temperature difference per unit distance).

Heat will flow down its thermal gradient, moving from a region of higher temperature to a region of lower temperature. The rate of heat flow is determined by several factors, including the area over which heat flow occurs, the initial temperature difference between the two

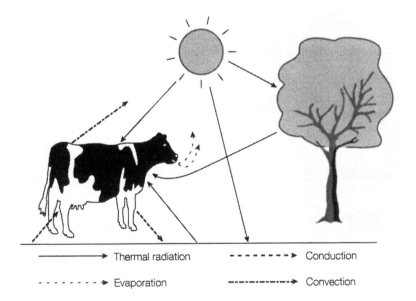

Thermal radiation   Conduction

Evaporation   Convection

Figure 6.1. The complexity of heat exchange between an animal and its environment. Arrows pointing towards the animal represent heat gain, arrows pointing away from the animal represent heat loss.

bodies and the thermal conductivity of the bodies involved. Thermal conductivity is a way of expressing how easily heat flows through a body. It will vary depending upon the nature of the material; for example, metals have a high thermal conductivity compared with plastic. Mammals, for example, have a low thermal conductivity which means that they are efficient insulators – they will resist heat flow. The explanation for this low conductivity (or, alternatively, high insulation) is that still air, which has an extremely low thermal conductivity, is trapped in biological tissue, such as fur and feathers. Thus, little heat will be lost to structures with which birds and mammals are in direct contact.

6.3.2 Convection

Convection is the transfer of heat by the movement of fluid. A fluid may be defined as either a liquid or a gas. It is one way in which heat loss from a solid may be increased. For example, heat can be lost from a solid to a gas. As the gas moves past the solid, heat is transferred from the solid to the gas. As the heated gas continues moving, it is replaced by cooler gas and more heat can be transferred from the solid to the gas. Convection is said to be forced if the movement of fluid occurs as a consequence of external forces, e.g. wind. The opposite of this – free convection – occurs in the absence of any external force.

6.3.3 Radiation

Radiation is the transfer of heat between two bodies that are not in direct contact with each other. Mathematically, it may be expressed by the following formula:

$$Q = \sigma \epsilon A T^4$$

where:
Q = heat radiated,
σ = Stefan–Boltzman constant,
ϵ = emissivity of the surface (this is identical to the absorptivity of a surface, i.e. how much incoming radiation of a particular wavelength is absorbed),
A = area from which radiative heat transfer is occurring, and
T^4 = temperature of the surface, raised to the power of 4.

What this equation is saying is that as the surface temperature of a body increases, radiative heat loss from that surface also increases.

All bodies at temperatures above 0°K emit electromagnetic radiation. The frequency and intensity of the radiation emitted depends upon the temperature of the body. The intensity of radiation increases (whilst its wavelength, which is related to frequency, decreases) as the temperature of the body increases, and at high temperatures radiation may be emitted that is in the visible region of the electromagnetic spectrum. This is the reason why materials (e.g. metals) glow when they are heated to extreme temperatures. In addition to emitting radiation, animals are also able to absorb radiation. This has important consequences in thermal physiology. So-called 'black bodies' absorb all types of electromagnetic radiation. The skin and fur of animals are considered to be black bodies, which means they will absorb solar radiation. This may represent an important source of heat gain. Dark coloured skin and fur will absorb more radiation than equivalent light-colored structures.

6.3.4 Evaporation

When water evaporates it moves from the liquid to the gas phase. This change of state requires a great deal of energy in the form of heat. This is known as the heat of evaporation and it is the reason why evaporation causes cooling. The amount of heat required for evaporation to occur depends upon the temperature at which it happens – as temperature increases, the amount of heat energy required to convert water from its liquid to its gaseous state decreases. Thus, evaporation is an important means by which heat may be lost; for example, when humans get hot they sweat, and when birds and dogs get hot they begin to pant. The fact that panting can be a significant source of heat loss also means that whenever air-breathing animals exhale heat is lost, since the gas breathed out is usually saturated with water vapor.

6.4 Ectotherms

As indicated previously, ectotherms are animals that are dependent upon external sources for heat gain; little of their body heat is obtained as a consequence of their overall metabolism. The problems of ectothermy are dependent upon whether animals are aquatic or terrestrial ectotherms.

6.4.1 Aquatic ectotherms

In many ways, the problems of aquatic ectotherms are minimized. This is because large bodies of water may provide a relatively stable environment in terms of temperature. This, together with other aspects of the physical chemistry of water, influences the temperature of animals which live in it. For example, there is no possibility of losing heat by evaporation, and heat exchange by radiation is also very much reduced since water is an effective absorber of infrared radiation. This means that the body temperature of an aquatic ectotherm is the same as the temperature of the water in which it lives.

Water is also a very effective heat sink – large amounts of heat may be dissipated to it. This means that all the metabolic heat generated by an organism may be dissipated into the water and there is a rapid loss of heat. In fish, for example, a major area of metabolic heat loss is via the gills. By definition, gills must be thin and well vascularized in order to meet the requirements for gas exchange, and it is precisely these attributes that allow heat to be rapidly lost from the blood as it passes through the gills. However, some fish, e.g. tuna and sharks, have been able to maintain a temperature differential through their bodies. This means that some regions are kept at a different temperature from other regions (*Figure 6.2*). This is advantageous to the fish because the elevated temperature means an increased rate of metabolic reactions in those parts. The regions which maintain a temperature greater than the surrounding water include the muscles used in swimming and parts of the digestive tract. The ability to maintain this temperature differential is due to the presence of heat exchangers in the fish, which are located between the tissue that is to be heated and the gills where the heat would otherwise be lost. The heat exchanger works on a countercurrent principle. Cooler arterial blood, which is arriving at the organs from the gills, runs alongside warmer venous blood which is leaving the organ concerned. Consequently, heat is transferred from the vein to the artery and the heat is contained within the organ (*Figure 6.3*). This arrangement is very effective and enables the swimming muscles to be maintained at a temperature some 12–15°C above that of the water in which the fish is swimming.

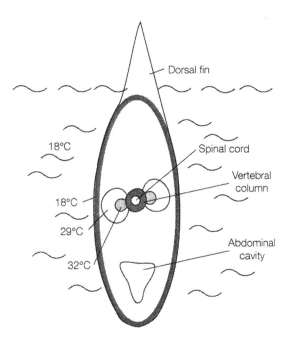

Figure 6.2. Cross section through a tuna fish, showing the temperature differential which is maintained through its body.

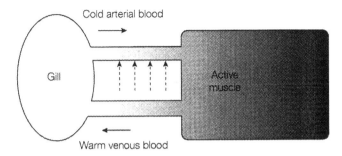

Cold arterial blood

Gill

Active muscle

Warm venous blood

Figure 6.3. Heat exchange in fish. Heat produced as a consequence of muscle metabolism is passed from warm venous blood into cooler arterial blood.

6.4.2 Terrestrial ectotherms

The thermoregulatory problems facing terrestrial ectotherms are rather different from those of aquatic ectotherms. The reason for this is that all of the processes described for heat exchange in Section 6.3 are now operating, at least, potentially. This means that terrestrial ectotherms have a greater chance of maintaining a body temperature which is different from that of the environment.

The most important way in which terrestrial ectotherms obtain heat is by the absorption of solar radiation. The absorption of solar radiation can be maximized by altering the color of the absorptive surface and the orientation of the animal towards the sun. It is well established that invertebrate terrestrial ectotherms alter their surface color to maximize solar radiation absorption – the darker the color, the greater the absorption. Such color changes, which are examples of physiological color change, are seen in some species of beetle and grasshoppers. This may have important implications for the animals concerned. Certainly it allows them to warm up earlier in the day, and may provide some sort of advantage in terms of obtaining food and water. The second way to maximize absorption of solar radiation is to change the orientation of the body towards the sun. For example, in order to maximize heat gain, locusts will position themselves at right angles to the sun so that they present a larger surface area to the sun's rays. Invertebrate terrestrial ectotherms must also regulate heat loss as body temperature cannot be allowed to rise unchecked. Heat loss by evaporation in many terrestrial invertebrates is minimal. Insects, for example, are covered by a waxy cuticle, the primary function of which is to prevent water loss. The simplest way for these animals to limit increases in body temperature, therefore, is to minimize absorption of solar radiation and heat gains by conduction from hot surfaces. This is done by altering the orientation of the body with respect to the sun, raising the body off the ground and climbing vegetation.

Responses similar to those seen in terrestrial invertebrate ectotherms are also seen in vertebrate terrestrial ectotherms. For example, lizards will bask in the sun to absorb solar radiation. They have the ability to alter melanin dispersal in their skin in order to maximize heat gain (increasing melanin dispersal in their skin will make it darker and therefore heat absorption will increase), and they will move into the shade to limit absorption. They are also able to alter bloodflow to the skin to maximize or minimize heat absorption by vasodilation or vasoconstriction respectively. Thus, there is a whole repertoire of physiological and behavioral means by which lizards may reach (and maintain) their optimal body temperature. This preferred body temperature, known as **eccritic temperature**, is usually between 35 and 40°C. Superimposed on this preferred body temperature are ranges over which body temperature is acceptable – the so-called **thermal tolerance range**. This is the range of temperatures over which the animal is still able to survive. Outside this range are the **critical minimum temperature** and **critical maximum temperature**. A body temperature below the critical minimum temperature or above the critical maximum temperature is not compatible with life. Those lizards which obtain the majority of their heat from the absorption of solar radiation are called heliotherms. This contrasts with a second group of lizards termed thigmotherms – these are nocturnal lizards, hence absorption of solar radiation is denied them. Therefore, in order to increase their body temperature they rely heavily upon conductive sources, such as lying on rocks and sand which have been warmed by the sun during the day.

6.4.3 Ectothermic adaptations to extreme temperatures

Adaptations to extreme cold.   Some ectotherms, e.g. fish and invertebrates, are able to live in extremely cold environments, experiencing temperatures which have the potential to freeze animal tissues and fluids. Various strategies have evolved in order to overcome this problem. Perhaps the easiest way is by the addition of extra solute molecules to their body fluids to increase their osmotic concentration (see Chapter 9). The addition of solutes to water lowers its freezing point so that it will freeze at a temperature below 0°C. The solutes added to body fluids to achieve this are either sugars, such as fructose, or sugar derivatives, such as glycerol. In some cases, the freezing point of body fluids may be lowered to nearly –20°C. Glycerol has another useful effect in that it protects membranes and enzymes from the consequences of denaturation at extremely low temperatures. A second strategy to cope with extreme temperatures is supercooling. Super-cooling is the phenomenon whereby the freezing point of water may be reduced to as low as –20 or –30°C. The body fluids of many ectotherms possess the ability to supercool to as low as –5°C. The actual mechanism of supercooling is unclear. One final option that has evolved that allows ectotherms to survive cold environments is the addition of antifreeze proteins to body fluids, such as those found in fish which inhabit Antarctic waters. The proteins, which are glycoproteins (protein conjugated with a carbohydrate), are polymeric molecules of a monomer consisting of a tripeptide linked to a galactose derivative (alanine-alanine-threonine-galactose derivative). There may be 50 or more of these monomers linked together. Several antifreeze proteins also exist which lack the carbohydrate moiety described above. These compounds work by lowering the freezing point of body fluids and are thought to work by inhibiting formation of ice crystals inside the animals. This is important since it prevents damage to the cell membrane.

Adaptations to extreme heat. Ectotherms exposed to high temperatures have two means by which they survive. The first of these is to increase the rate of cooling by evaporation (sweating). In many cases, this does not present a problem since many ectotherms are moist skinned. These animals simply increase cutaneous water loss. This is more of a problem for ectothermic animals which have a covering designed to minimize water loss, e.g. reptiles and insects. In this case, water is lost from the respiratory tract and heat is also lost. The second possible solution is for the animal's metabolic machinery to become adapted to working at high temperatures. By elevating their upper lethal temperature in this way, these animals are able to exist at temperatures which would have a fatal effect on most other animals. In most cases, the lethal temperature is within the range 30–45°C. This contrasts with organisms, such as some bacteria, which have been found living in environments just a few degrees below 100°C.

6.5 Endotherms

In contrast to ectotherms, endotherms are animals whose body temperature is derived from internal heat production, which is a by-product of cellular metabolism, and whose body temperature remains relatively constant, irrespective of changes in the temperature of the external environment. Endothermy is generally considered to be the province of birds and mammals. However, as described in Section 6.4.1, some fish (e.g. tuna) are able to maintain a temperature in some regions of their body which is significantly different from that of the external environment. Such responses are also observed in some reptiles; for example, the female python maintains an internal body temperature some 5°C above that of the external environment. This is particularly important in the brooding and development of eggs.

Birds and mammals are classic endotherms. There is a wide variation in actual body temperatures (e.g. the duck-billed platypus has a body temperature of about 30°C, whilst woodpeckers have a body temperature of about 42°C). In all cases, the body temperature is maintained at its desired level by reaching a balance between heat production and heat loss. The different mechanisms of heat production and heat loss are summarized in *Figure 6.4*.

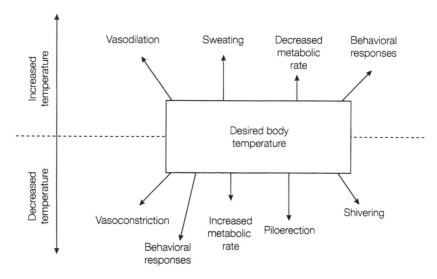

Figure 6.4. Summary of some the factors which contribute to the maintenance of a constant body temperature in endotherms.

6.5.1 Mechanisms of heat production

The principal mechanism for increasing the production of heat is to increase metabolic heat production in skeletal muscle. Heat is released as a by-product of intracellular metabolism in muscle cells as a consequence of the contractile process. This means that in order to generate heat the muscle must be contracting. In some instances, such muscular movement may be voluntary – a good example would be that of humans stamping their feet and rubbing their hands when cold. Generally, though, this increased muscular activity is involuntary and is termed shivering. Shivering is asynchronous muscle contraction; no purposeful movement is achieved. It may increase heat production by up to five-times that of resting heat production. Shivering has been demonstrated in many mammals and birds and has also been observed in some insects such as bees, where it is thought to keep the flight muscles warm.

There are also a number of other, more biochemical means by which heat production may be increased in endotherms. These mechanisms are collectively termed nonshivering thermogenesis. One mechanism involves brown adipose tissue (BAT), sometimes called brown fat, which has only been identified in the eutherian (placental) mammals. Brown fat, which is different from normal (white) adipose tissue, is packed with fat droplets and is extensively innervated by the sympathetic autonomic nervous system. When BAT is stimulated, the fat is metabolized within the mitochondria of the fat cells and heat is produced. The disadvantage of heat production in this manner is that the oxygen requirements of the animal are immediately increased in order to metabolize the fat.

Metabolic heat production may also be increased by the activity of the thyroid hormones thyroxine and triiodothyronine. The release of these hormones is under the control of the anterior pituitary gland, as shown in *Figure 6.5*. Both hormones enter cells where most of the thyroxine is converted to triiodothyronine. Once this conversion has occurred, triiodothyronine produces several effects, including stimulation of Na^+/K^+-ATPase and several enzymes involved in the metabolism of glucose. Overall, the effect is an increase in metabolism and, therefore, an increase in heat output. In order to sustain this increased metabolism, O_2 uptake must also increase and this is achieved by a triiodothyronine stimulated increase in the rate of breathing. The release of the thyroid hormones is regulated by a negative feedback loop. When the levels of thyroxine and triiodothyronine become elevated, they inhibit further release of thyrotropin releasing hormone from the hypothalamus and thyrotropin-stimulating hormone from the anterior pituitary gland, thus limiting the release of the thyroid hormones. Many endocrine secretions are controlled by negative feedback.

Finally, there are a number of other mechanisms by which the body temperature of an endotherm may increase. Perhaps the simplest is by absorption of solar radiation (like ectotherms), but other methods include erection of body surface hairs which traps a layer of still air next to the skin and thereby reduces convective heat loss, and reducing blood flow to peripheral organs by vasoconstriction. This may be further aided by the use of heat exchangers. There are also a variety of behavioral responses that animals can utilize to increase body

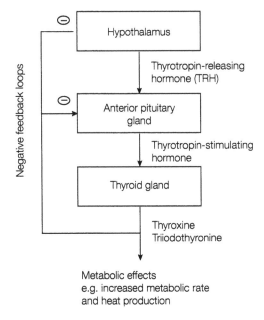

Figure 6.5. The control of the secretions of the thyroid glands.

temperature. For example, humans may add extra layers of clothing prior to entering a cold environment.

6.5.2 Mechanisms of heat loss

Heat which is lost is dissipated to the environment. One way in which this is achieved is to divert blood to the periphery of the animal by vasodilation of peripheral blood vessels. Humans often appear red and flushed when they are hot and this is due to vasodilation of blood vessels in the skin. Vasodilation and vasoconstriction are under the control of the autonomic nervous system. Vasodilation occurs when there is a reduction in the activity of nerves which promote vasoconstriction and an increase in the activity of those which cause vasodilation (the former being much more numerous). A second way by which heat may be lost is via evaporation of water. In many mammals, heat loss is achieved by the evaporation of sweat from the skin. However, some mammals (e.g. dogs and all birds) do not possess sweat glands and are unable to sweat. In this situation, water is lost from the respiratory tract by panting. Panting is done in such a manner (by taking rapid, shallow breaths) that disturbance to the concentration of blood gases is minimized. For example, rapid, deep breathing would result in the 'blowing off' of CO_2, and its plasma concentration would drop to a level that would result in the removal of a major stimulus to breathe. Some animals achieve the cooling effect of water evaporation by other means. The kangaroo, for example, will lick its fur, and the evaporation of the saliva will have a cooling effect.

6.5.3 Endothermic adaptations to extreme temperatures

Adaptations to extreme cold. The fact that endotherms live in cold, hostile environments suggests that they possess some specialized mechanisms, physiological or otherwise, which allow them to do so. The simplest way to cope with an extremely cold environment is to tolerate it. This can be achieved by so-called regional heterothermy, the maintenance of different parts of the body at different temperatures. A good example of animals displaying regional heterothermy are arctic birds and mammals, which, whilst maintaining core temperatures at around 38°C, can maintain their feet at a temperature of around 3°C. In those animals displaying regional heterothermy, all other aspects of the animal's physiology are normal (the nervous system in the foot operates perfectly well at 3°C). This implies adaptation at the cellular and molecular level. Perhaps the most extreme response seen to low environmental temperatures is **hibernation**, or torpor. Hibernation may be defined as a reduction in body temperature (i.e. hypothermia) accompanied by a corresponding decrease in metabolic rate, heart rate, respiratory rate, and so on. Obviously for some animals, food is scarce during the winter months, and given that it is the metabolic breakdown of food which generates heat, the ability to maintain an adequate core temperature during this period would be severely compromised. The period of hibernation may last from a few hours through to days, weeks or months. Termination of the hibernating state is achieved by spontaneous arousal, with a rapid increase in metabolic rate and body temperature, which returns to its normal value. Some animals (e.g. bats) will enter into a hibernating state on a daily basis. Once again, this is related to their lifestyle which dictates an exceedingly high metabolic rate that they are unable to maintain continuously.

Adaptations to extreme heat. Endotherms which tend to maintain core body temperature of around 37–42°C are able to withstand hot environments. The ability to do so depends on how hot the environment is, the size of the animal and how well insulated it is. This means that a large, well insulated endotherm in a given hot environment will be heat-stressed before a smaller, less well insulated endotherm. There are several strategies that endotherms utilize in order to maintain their core body temperature within normal limits in the face of an excessive increase in external temperature. The simplest way to increase heat loss is to increase water loss by evaporation. Man and other mammals achieve this by increased sweating. However, as mentioned previously, some mammals, such as dogs, and all birds are incapable of sweating. Therefore, water loss must come from the respiratory tract and this can be achieved by panting. Birds display a phenomenon called gular fluttering, rapid movement of the throat region which increases evaporation and, hence, increases heat loss. In the majority of cases, the respiratory loss of water is not accompanied by any change in blood gas concentrations or any disruption to the animal's acid–base balance.

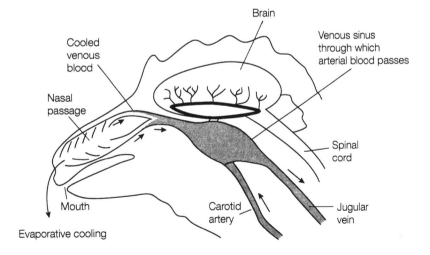

Figure 6.6. Warm arterial blood passes in close proximity to cooler venous blood in the carotid. Heat is transferred from arterial to venous blood and the brain is supplied with cool blood. Redrawn from Hardy, R.N., *Temperature and Animal Life*, 2nd edn, 1979, with permission from Edward Arnold.

A potentially more dangerous strategy is for the animal to become hyperthermic by transiently retaining metabolic heat and raising its body temperature. A good example of this is the desert camel. There are many advantages of temporary hyperthermia, not least of which is the reduction in the amount of water which must be lost to produce cooling. In hot environments water may be at a premium, so it is advantageous to be able to limit its loss. However, hyperthermia also presents animals with problems. For instance, the elevated body temperature may not be equally tolerated by all organs and the possibility exists that damage could occur to some organs or tissues within the animal. One such organ is the brain. In order to perfuse the brain with blood at the correct temperature, many such mammals have heat exchangers in the brain, whereby the heat in arterial blood flowing into the brain is transferred into venous blood leaving the brain (*Figure 6.6*).

6.6 Control of body temperature in endotherms

Like all control systems, the control of temperature is an example of homeostasis. As such, it has the components of any homeostatic system – receptors, comparators, effectors and of course neural elements linking them together.

The receptors which monitor temperature are called **thermoreceptors**. Generally, there are two types of thermoreceptor: those which increase their action potential production in response to an increase in temperature and those which increase action potential production in response to a decrease in temperature. These represent 'hot' and 'cold' thermoreceptors, respectively. Thermoreceptors are found in two locations – the peripheral sensors in the skin and the central receptors in the hypothalamic region of the brain. The first group monitor peripheral changes in the temperature of the animal, whilst the second group monitor changes in the core temperature. It is necessary to have receptors situated in these two locations. Consider what would happen if only central receptors were present: any temperature changes which occurred in the periphery would have to be transmitted through the animal before the central receptors were stimulated. If there was a significant change in temperature at the periphery, considerable tissue damage could be done before this temperature change was detected by the central thermoreceptors. The hypothalamus is also the site of the set point of temperature, the desired temperature of the body. For humans, this is 37°C, whilst for some birds (e.g. parrots) it may be as high as 42°C. Thus, the hypothalamus acts as a thermostat for the body.

On the basis of 'selective' lesioning experiments (the destruction of discrete regions of the hypothalamus), it has been proposed that the different regions of the hypothalamus perform different thermoregulatory functions. It has been suggested that the anterior hypothalamus is a 'heat loss' centre since mammals which have had it destroyed find it difficult to thermoregulate when heat stressed. There is also a posterior 'heat gain' centre. Damage here results in animals losing the ability to generate extra heat when exposed to the cold or to maintain normal heat production. The neurophysiological explanation of how the hypothalamus acts as a thermostat is far from clear. However, it is known that on the basis of neural input from thermoreceptors and its function as a thermostat it is able to influence the function of various effectors in order to maintain a constant body temperature.

6.7 A comparison of ectothermy with endothermy

On first consideration, it might appear that animals have nothing to gain from an ectothermic way of life. In terms of cost to the animal, though, this is not necessarily so. If an animal is ectothermic, in metabolic terms, it costs the animal nothing in order to maintain some sort of appropriate body temperature for at least part of the day. The downside to this is that as all reactions are governed by temperature – neuronal function and digestive processes to name but two – when body temperature decreases so does the activity of many organ systems. Thus, such animals are vulnerable to predation, for example, when their body temperature is reduced. In contrast, endothermy provides a relatively constant body temperature. This means that metabolic reactions may be governed by enzymes which display optimum activity at body temperature. Thus, endotherms are not prone to the consequences of changes in temperature that ectotherms are subjected to. It would be impossible for ectotherms to evolve enzymes governing metabolic reactions which had a wide optimum temperature. This means, then, that endotherms are capable of utilizing colder environments more successfully than ectotherms. The downside to an endothermic way of life is that animals require a greater O_2 uptake, so metabolic rates must be higher.

Further reading

Hardy, R. N. (1979) *Temperature and Animal Life.* Edward Arnold, London.

Circulatory systems

7.1 Functions of circulatory systems

The circulatory system of animals ensures the rapid, bulk movement of substances around the body. Such a system is required when the movement of substances by simple diffusion is inadequate, such as the exchange of gases or nutrients across the general body surface area directly into cells. It should be noted that even within individual cells there is an organized system for the transport of substances. For example, consider neurons where substances may be made in the cell body but are required at the axon terminal which may be several centimeters away – this involves the process of axonal transport described in Chapter 2. In some animals, e.g. vertebrates, the medium for the transport of substances is blood or vascular fluid which is contained within blood vessels. This is known as **closed circulation**. In animals, such as insects, which have an **open circulation**, a system in which the blood does not make a complete circuit around the animal within blood vessels, the circulating fluid is referred to as **hemolymph** and it flows out of blood vessels into a body cavity or **hemocoel**, bathing the tissues directly (see Section 7.4.1). The list below indicates some of the substances which may be transported in blood.

- The respiratory gases O_2 and CO_2.
- Nutrients, which are transported, for example, from the gastro-intestinal tract to storage organs and from storage organs to sites of utilization.
- Waste products, e.g. urea is transported from liver to kidneys, and CO_2 is transported from the tissues to the gas exchange organs.
- Specialized blood cells, e.g. white blood cells (leukocytes) which participate in immune and defence reactions and platelets (thrombocytes) which participate in hemostasis (blood clotting).
- Hormones. In many cases the hormone is carried attached to some kind of transport molecule, usually a protein. This is the mechanism by which the steroid hormones are transported.
- Heat may be transferred between organism and environment as it flows through the vascular beds of the skin.

In many cases, the substance that is being transported is simply dissolved in the vascular fluid, although there are some important exceptions to this, such as the transport of oxygen which is performed with the aid of specialized transport molecules, which will be discussed later. All cardiovascular systems require three basic components – a circulatory fluid, a heart to generate pressure and, therefore, pump the circulatory fluid, and a system of tubes through which the fluid can circulate, although this latter component is not present in all circulatory systems. Each of these will be looked at in turn.

7.2 The composition of blood

Blood is the substance by which various substances are transported around the bodies of animals. The blood may be contained within specific vessels or it may flow freely between cells of the body, in which case it is known as hemolymph. In general, blood consists of cellular elements (red and white blood cells) within a fluid matrix, the plasma.

7.2.1 Composition of plasma

Plasma is the fluid matrix in which the cells of the blood are suspended. It is basically an aqueous solvent in which varying amounts of ions and organic molecules, including proteins, are dissolved. The composition of plasma is generally very different from that of intracellular fluid – there is a high concentration of sodium and low concentration of potassium in blood, whilst the opposite is true of intracellular fluids. There is also a wide variation in the protein content of the plasma. This has important consequences for the osmotic pressure of the plasma (see Chapter 8). Given that proteins are large molecules which are impermeable to cell membranes, they are 'trapped' within the plasma. The greater the amount of protein, the greater the osmotic pressure. The osmotic pressure generated by proteins is called the **colloid osmotic pressure**. The significance of this, and any other osmotic pressure, is that it influences the movement of water across cell membranes. When the concentration of proteins in plasma is viewed across the animal kingdom, it can be seen that there is a huge variation in concentration between species. For example, the concentration of protein in jellyfish is about 0.5 g l^{-1}, whilst in some vertebrates it may be as high as 80 g l^{-1}. This may well be related to the osmotic 'lifestyle' of each of these animals. Jellyfish are **osmoconformers**, which means that their body fluids simply assume the same osmolarity of their external environment. Vertebrates, on the other hand, are **osmoregulators**, maintaining the osmolarity of their body fluids within relatively narrow limits. As mentioned previously, the colloid osmotic pressure is one way of influencing water movement and, therefore, body fluid osmolarity within animals.

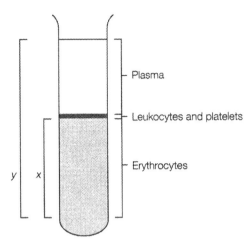

Figure 7.1. A sample of blood after centrifugation. The proportion of blood volume occupied by erythrocytes is known as the hematocrit. It is calculated as $x/y \times 100$.

7.2.2 Blood cells

Within the fluid matrix of blood are the cellular elements. The percentage of the total blood volume which is occupied by cells is known as the **hematocrit** (*Figure 7.1*). The cells within the circulatory system are given the general name of hemocytes and they have several functions. These include the transport of oxygen and, to a very limited extent, carbon dioxide, defense (e.g. some are motile and are able to absorb and digest foreign particles) and hemostasis, by aggregating around holes or tears in blood vessels to form a seal, thus preventing blood loss temporarily. Generally, those cells which transport oxygen are called **erythrocytes**, those involved in defence are called **leukocytes** and those involved in hemostasis are called **thrombocytes**. *Figure 7.2* shows the appearance of erythrocytes from a variety of animals.

The types of cells actually found in the blood of animals varies tremendously between the different phyla. For example, echinoderms (i.e.

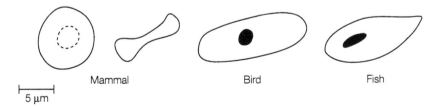

Figure 7.2. The appearance of erythrocytes in mammals, birds and fish. There is tremendous variation in the size of erythrocytes. Mammalian erythrocytes are without nuclei, whilst other vertebrates generally have nucleated erythrocytes.

starfish, sea cucumbers) have erythrocytes and a variety of other cells called coelomocytes. The coelom is a body cavity which is completely lined by tissue of mesodermal origin (*Figure 7.3*). There are several different types of coelomocytes (e.g. amoebocytes), all of which perform functions similar to leukocytes in higher animals, participating in immune responses, blood clotting and so on. Many animals (e.g. annelid worms, arthropods) have only cells which are involved in the defensive reactions previously described. All vertebrates have both erythrocytes and leukocytes, of which several sorts may be distinguished (*Figure 7.4*).

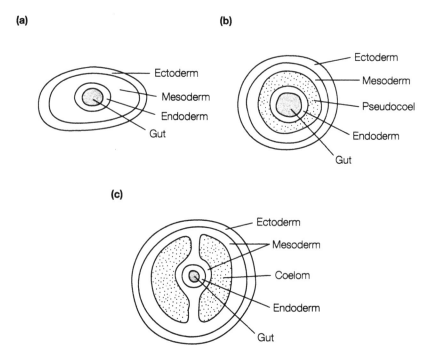

Figure 7.3. (a) Acoelomate animals possess no body cavity (e.g. platyhelminths); (b) pseudocoelomates possess a cavity but it is not lined entirely by mesoderm (e.g. nematodes); (c) coelomates have a true coelom lined by mesodermal tissue (e.g. annelids).

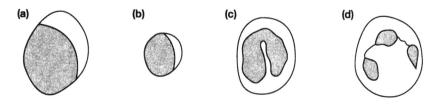

Figure 7.4. Vertebrate leukocytes. There are many subtypes, all of which originate from a common precursor cell. (a) Large lymphocyte; (b) small lymphocyte; (c) monocyte; (d) granulocyte. (Not drawn to scale).

7.3 The heart

The role of the heart in circulatory systems is two-fold. Its first function is pumping fluid around the circulatory system. This is achieved by alternate contraction and relaxation of cardiac muscle, thus providing a pressure gradient forcing blood out of the heart and around the body of the animal – the heart is a pressure pump. The second role of the heart is to exert some control over the circulatory system, by alteration to the rate of beating and the force of contraction, for example.

Before discussing the structure and function of the heart in different animals, it should be noted that some animals achieve movement of blood without the need for a specialized organ such as the heart. It is possible that blood may be moved along blood vessels when the vessel is surrounded by blocks of skeletal muscle (*Figure 7.5*); when the muscle contracts, blood is squeezed along the vessel. This type of arrangement is found in, for example, nematode worms, echinoderms and vertebrates. In the first two examples, movement of the body wall muscle results in the movement of fluid in the fluid-filled cavities which form the pseudocoelom and coelom, respectively. In the latter example, such an arrangement is important for returning blood to the heart – as skeletal muscle contracts, blood is forced back to the heart. On relaxation of the muscle, blood return is prevented by the presence of valves in the veins.

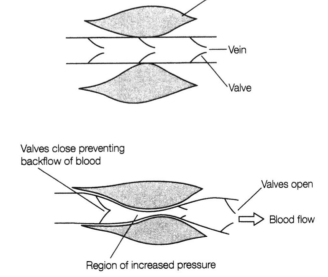

Figure 7.5. During muscle contraction, a region of high pressure is generated. This is the driving force for blood movement. The presence of valves in the blood vessel ensures that blood flow is unidirectional.

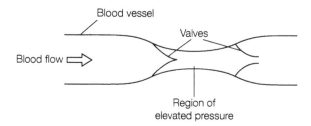

Figure 7.6. The tubular heart. Peristaltic-like contraction of the blood vessel forces blood through the system. The region of elevated pressure is the driving force for the movement of blood. Backflow is prevented by valves in the wall of the blood vessel.

Perhaps the simplest heart proper, is the so-called tubular heart. In essence, this is a tubular structure which contracts in a manner similar to peristaltic movements, forcing blood along the tube. This type of heart is seen in many insects. In order to ensure a one-way flow of blood, the heart may be valved (*Figure 7.6*).

More complex than the tubular heart is the chambered heart (*Figure 7.7*). In this case, the heart consists of a number of neighboring chambers which function in a coordinated fashion to propel blood around the circulatory system. Chambered hearts are found in molluscs and in vertebrates, but the number of chambers varies from species to species. In general, the chambers of such hearts may be classified into

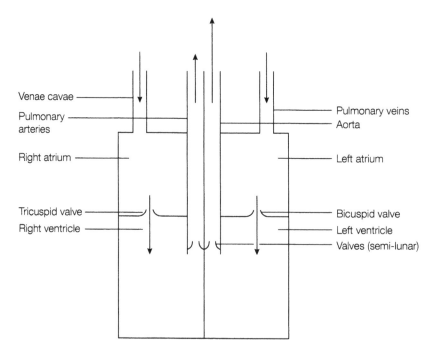

Figure 7.7. Stylized structure of the mammalian heart.

atria and ventricles. The atria are essentially collection chambers which pass the blood into the ventricles. The ventricles are surrounded by a thick layer of cardiac muscle, and it is the contraction of this muscle that creates the pressure which is the driving force for the movement of blood around the animal.

Hearts may be classified into two types on the basis of how the heart-beat is initiated: neurogenic and myogenic. Neurogenic hearts depend upon extrinsic neural innervation to initiate contraction. If this innervation is removed, then the heart ceases to beat. Neurogenic hearts are found in crustaceans, spiders and some insects. In crustaceans the heartbeat is regulated by the neural activity (i.e. action potentials) arising in the cardiac ganglion. The ganglion acts as a pacemaker, originating periodic trains of action potentials which then pass to the heart where they cause the cardiac muscle to contract. Superimposed upon this are other nerves, both inhibitory and excitatory, which modulate the activity of the cardiac ganglion. The identity of the neurotransmitter released by these cardiac neurons is unclear. Generally, though, the neurotransmitters dopamine, serotonin (5-HT) and noradrenaline (norepinephrine) have been shown to have stimulatory effects on the heart, whilst acetylcholine and γ-aminobutyric acid (GABA) are known to exert inhibitory effects.

Myogenic hearts are those which display spontaneous contractile activity. They are found in the molluscs and vertebrates, for example. The contraction is a consequence of the spontaneous neural discharge of a particular region of the heart. This region is known as the **pacemaker** and it is a small region of modified cardiac muscle, the cell membrane of which has an unstable resting membrane potential. Therefore, the resting membrane potential regularly 'drifts' towards threshold. Each time threshold is reached, an action potential is generated and heartbeat is initiated. The depolarization which originates here travels to the rest of the cardiac muscle, and this causes contraction and, therefore, pumping of blood. In gastropods, it is sometimes difficult to distinguish the pacemaker region, although this is not the case in vertebrates, where the pacemaker is known as the sino-atrial node. In myogenic hearts, all regions of the heart have the ability to spontaneously depolarize, and in situations where the primary pacemaker region fails, another region of the heart will assume pacemaker activity. Although myogenic hearts have an inherent contractility, their action is modified by both neural and endocrine influences. For example, acetylcholine and noradrenaline are neurotransmitters which have their sites of action in the pacemaker region of the heart. The activity of the heart is decreased by acetylcholine and increased by noradrenaline in both molluscs and vertebrates. Acetylcholine decreases the heart rate by altering the ionic events which are responsible for depolarization at the pacemaker. In this case, potassium efflux from the pacemaker cells increases which results in hyperpolarization, and calcium influx is inhibited, preventing the development of action potentials in this region of the heart. Noradrenaline, on the other hand, increases heart rate. Again, this is due to effects at the pacemaker region. The effect of noradrenaline is to increase calcium influx into cells of the pacemaker region, thus reducing the time for depolarization to occur and increasing the heart rate.

7.4 Types of circulatory system

7.4.1 Open circulatory systems

An open circulatory system is one in which the blood vessels do not form a complete circuit around the body. Instead, the blood leaves the vessels and flows over and between the tissues of the animal before returning to the heart. This type of system is found in arthropods and many molluscs. The typical arrangement found in the arthropods is shown in *Figure 7.8*. The heart provides the force which drives blood around the animal. The heart itself has a number of openings called ostia which allow blood to re-enter the heart. In many cases, the relaxation of the heart actively 'sucks' blood back into it due to the generation of negative pressures within the heart chambers. Although relatively simple in design, there are a number of disadvantages to the operation of open circulatory systems. One of these is that the system operates at low pressure, by virtue of the fact that a small volume of blood is ejected from the heart into a relatively large cavity. Given that circulation of the blood is pressure-driven, blood delivery to the tissues may be slow. This imposes limits on the rate of delivery of nutrients to cells and may, in turn, limit the metabolic activity of the animal concerned. Another disadvantage is that the animal has no means of regulating blood flow to different organs. It is impossible to ensure that blood goes to any individual organ at any particular time – rather, blood flow to particular organs 'just happens'.

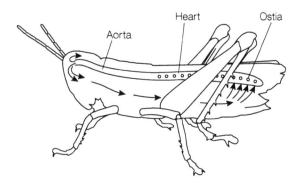

Figure 7.8. Arrangement of the arthropod circulatory system. Arrows indicate the direction of blood flow. The aorta itself may be contractile and this aids the flow of blood. The aorta branches to supply the tissues of the body with blood.

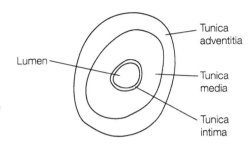

Figure 7.9. The basic organization of arteries and veins. Capillaries consist of the tunica intima, or endothelium, only. This is continuous throughout an animal's circulatory system.

7.4.2 Closed circulatory systems

In this arrangement, the blood of an animal is contained exclusively within a series of blood vessels. Closed circulatory systems are found in vertebrates, cephalopod molluscs and echinoderms. Such an arrangement confers a number of advantages when compared with the open circulatory system. The blood, which is moved by the pumping activity of the heart, is maintained constantly at high pressure, with the result that blood exits and returns to the heart rapidly. This ensures a rapid delivery of nutrients to metabolizing cells and, since blood is contained within vessels, it can be delivered directly to cells. Perhaps equally important is the fact that the distribution of bloodflow around an animal can be altered. For example, during periods of increased metabolic activity such as exercise, vertebrates have the ability to increase blood supply to active regions, such as muscle, at the expense of inactive regions, such as the gastrointestinal system.

Blood vessels of vertebrates. The closed circulatory system of vertebrates is made up of three types of blood vessel: arteries, capillaries and veins. Arteries and veins consist of three concentric circular layers of tissues. These layers (working from the lumen outwards) are the tunica intima (or endothelium), the tunica media and the tunica adventitia (*Figure 7.9*). Capillaries consist of the tunica intima only. The precise composition of each of these layers varies between vessels and is summarized below in *Table 7.1*.

The precise composition of a blood vessel is directly related to the function that the blood vessel performs.

Table 7.1. The composition of vertebrate blood vessels. The number of + signs indicates the relative abundance of each component

Component	Type of blood vessel				
	Artery	Arteriole	Capillary	Venule	Vein
Endothelium	Present	Present	Present	Present	Present
Smooth muscle	+++	++++	Absent	+	+++
Elastic fibers	++++	++	Absent	+	++
Connective tissue	+++	++	Absent	+	++

(i)   Arteries convey blood away from the heart. The blood at this
       point in the cardiovascular system is at a considerable pressure.
       Therefore, the wall of the vessel must be thick and strong enough
       to withstand this pressure as blood is delivered into them from
       the heart. The elastic fibers are particularly important in ensuring
       that blood delivery to the capillaries is constant. Remember that
       blood delivery from the heart is intermittent – when the heart
       contracts blood is pushed into the vascular system. As blood enters
       the large arteries it causes their walls to stretch, storing elastic
       strain energy. When the heart relaxes, the walls of the large arteries
       elastically recoil to their original size. In doing so, they release
       most of the stored energy forcing the blood to move further around
       the circulatory system. In doing this they convert the intermittent,
       pulsatile flow of blood from the heart into a continuous flow at
       the level of the capillaries. The pressures in the arteries when
       the heart is contracting and relaxing are termed the systolic and
       diastolic pressure, respectively. These pressures are above normal
       atmospheric pressure (760 mmHg) and vary from species to
       species, and are recorded as systolic/diastolic. In humans, for
       example, blood pressure is typically 120/80 mmHg (although it
       varies with age, disease and other factors), whilst in fish it is only
       30/20 mmHg.

(ii)   Arterioles are small arteries. Their walls contain large amounts of
       smooth muscle, which is not under voluntary control. It was stated
       earlier that an advantage of closed circulatory systems was that
       blood flow to different organs could be regulated. This is the role
       of the arterioles. By altering the degree of contraction of this
       smooth muscle, it is possible to regulate the flow of blood into a
       particular organ. The contraction of this muscle is called vaso-
       constriction whilst its relaxation is called vasodilation. The state
       of contraction of this muscle is controlled by the autonomic
       nervous system. However, control by the nervous system can
       be overridden by local factors. A good example of this is found
       in exercising muscle. The metabolic waste products of exercising
       muscle (e.g. lactate, CO_2 and H^+) diffuse to the arterioles and cause
       them to vasodilate. This ensures that the blood flow into the
       muscle is increased, which maximizes delivery of oxygen and
       other essential nutrients. This is an example of a local homeostatic
       response.

(iii) Capillaries are the smallest vessels in the circulatory system. They
       are the sites of exchange of gases, nutrients and other substances
       between blood and cells. This function is aided by capillaries being
       only one cell thick and by virtue of the fact that they have an
       enormous surface area.

(iv)   Venules (small veins connected to capillaries) and veins return
       blood to the heart. The pressure within the vascular system has
       fallen considerably by now, hence the walls of these vessels are
       much thinner than the walls of arteries. The majority of veins
       are valved to ensure a one-way flow of blood back to the heart.

Arrangement of closed circulatory patterns. Closed circulation systems can be arranged in different ways. This really relates to the way in which the heart is arranged and the way in which blood completes a full circuit around the animal. Perhaps the simplest arrangement is seen in fish (*Figure 7.10*). In this situation, blood leaves the heart via the ventricle, passes to the gills where it is oxygenated and from there goes to the rest of the body tissues before returning to the atrium of the heart, to begin the cycle again. This type of circulatory pattern is called a single circulation. The major disadvantage of such an arrangement is that pressure is lost as blood passes through the gills. As a result of this, blood flow from the gills to the rest of the tissues is sluggish, because the pressure gradients that are the driving force for the movement of blood are reduced. In contrast to this arrangement is that seen in mammals (*Figure 7.11*). In this case, the heart has four chambers, two upper atria and two lower ventricles. This means that for blood to make a complete passage around the body it passes through the heart twice. Blood leaves the right ventricle and passes to the lungs where it is oxygenated. It returns to the left atrium and passes to the left ventricle from where it is ejected and pumped to all the other tissues of the body. The deoxygenated blood returns to the right atrium, passes to the right ventricle and the whole cycle begins again. Thus, the heart in mammals can be thought of as two pumps: one dealing with the flow of blood to the lungs, the other dealing with the flow of blood to the rest of the body. This arrangement is known as double circulation. Double circulation overcomes the problem of pressure loss as blood passes through the gas exchange organs, which is seen in animals with a single circulation. An alternative solution to this problem has been developed by the cephalopod

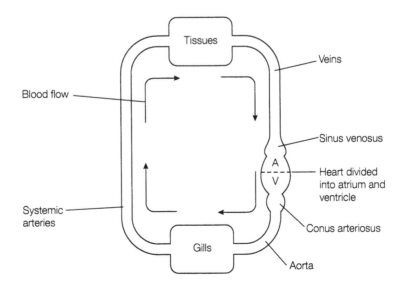

Figure 7.10. The single circulation system of fish.

molluscs, such as octopods (*Figure 7.12*). In this case, the animal has more than one heart, the additional hearts being called branchial hearts. As blood passes through the gills it loses pressure. The pressure of the oxygenated blood returning to the heart would therefore be very low.

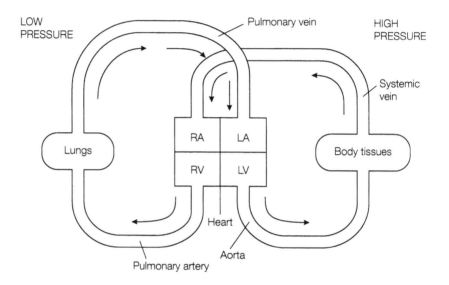

Figure 7.11. The double circulation system in mammals. Arrows indicate the direction of blood flow. The right side of the heart pumps blood to the lungs (pulmonary circulation), whilst the left side pumps blood to all other tissues of the body (systemic circulation). RA, right atrium; RV, right ventricle; LA, left atrium; LV, left ventricle.

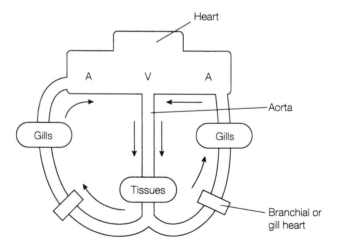

Figure 7.12. The circulatory system of the octopus. The heart is separated into two atria and a single ventricle. The presence of branchial hearts aids blood flow.

However, contraction of the branchial heart increases the pressure of the blood as it enters the gills, and the pressure of the oxygenated blood leaving the gills is therefore much higher than it would be if the branchial heart was absent.

7.5 Transport of oxygen

The transport of O_2 in the blood can occur in one of two ways: it can be carried in simple solution (i.e. dissolved in blood plasma) or it may be carried in conjunction with a respiratory pigment, a specialized compound which displays reversible binding of O_2 molecules. Many invertebrates transport O_2 dissolved in blood. They generally have low metabolic rates, and, therefore, their O_2 requirements are lower. However, in higher invertebrates and vertebrates, metabolic rates are higher and, therefore, the O_2 requirements of tissues are greater. Thus, the role of the respiratory pigment is to increase the amount of O_2 which can be transported in blood. In mammals, the presence of the respiratory pigment hemoglobin increases the O_2 carrying capacity of blood 20-fold, compared with the amount that can be carried in simple solution (i.e. dissolved in plasma). Hemaglobin allows 20 ml of O_2 to be carried per 100 ml of blood.

7.5.1 Respiratory pigments

All of the known respiratory pigments are conjugated proteins – they consist of a protein coupled with a nonprotein component. The nonprotein component is a metal, usually iron or copper, together with, in some cases, other components. There are four substances which are known to function as respiratory pigments. These are hemocyanin, chlorocruorin, hemerythrin and hemoglobin (*Table 7.2*). In those animals in which the respiratory pigment is free in solution, the pigment usually exists as an aggregate of many individual pigment molecules. This has the advantage of minimizing the osmotic problems that would ordinarily be caused by having a large number of molecules dissolved in the plasma, whilst at the same time ensuring that

Table 7.2. The principle characteristics of respiratory pigments

Pigment	Conjugate	Location	Occurrence
Hemocyanin	Cu^{2+}	Free in solution	Crabs, lobsters, cephalopods, snails
Chlorocruorin	Fe^{2+}	Free in solution	Four families of polycheate worms
Hemerythrin	Fe^{2+}	Free in solution or in blood cells	Sipunculids, brachiopods, some annelids
Hemoglobin	Fe^{2+}	Free in solution or in blood cells	Some flatworms, nematodes and annelids, some arthropods, some molluscs, virtually all vertebrates

O_2 transport and delivery to the cells is not compromised. The packaging of respiratory pigments into blood cells confers advantages to the animal. Again, it limits the osmotic problems associated with large numbers of molecules dissolved in plasma. By enclosing the pigment into cells the chemical environment of the cell can be specialized to optimize O_2 transport.

7.5.2 *Transport of O_2 by hemoglobin*

The function of hemoglobin in O_2 transport is probably the best understood of all the respiratory pigments. The structure shown in *Figure 7.13* is the monomer or basic building block of hemoglobin. In mammalian blood, for example, four of these subunits join together to make the final hemoglobin molecule, which is a tetramer. The role of hemoglobin is to bind reversibly with O_2 at the site of gas exchange structures and deliver it to the tissues of the animal. In all cases, the movement of O_2 from lung to blood and then from blood to cell occurs by the process of diffusion, with O_2 moving down its concentration gradient. The combination of O_2 with hemoglobin can be represented as follows:

$$4Hb + 4O_2 \rightarrow 4HbO_2$$

It should be remembered that the process is not one of chemical oxidation. Rather there is an association between the Fe^{2+} at the center of the heme group and a molecule of O_2.

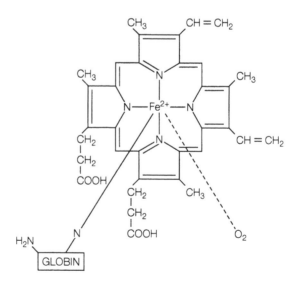

Figure 7.13. The structure of hemoglobin. Different types of globin are found in fetal and adult mammals. The globin molecule is made up of between 140 and 146 amino acid residues.

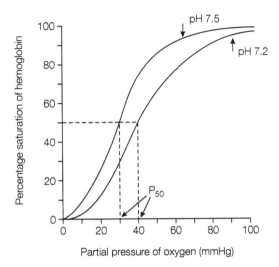

Figure 7.14. The oxygen–hemoglobin dissociation curve showing the effect of pH. As the pH decreases, hemoglobin has less affinity for O_2 and more O_2 is released into tissue. This is known as the Bohr effect.

The binding of O_2 to hemoglobin is best shown by the oxygen–hemoglobin dissociation curve (*Figure 7.14*). Experimentally, a sample of blood in which no O_2 is bound is exposed to increasingly higher concentrations of O_2 (i.e. the pO_2 is increased), and the degree of oxygen bound (its saturation) is measured. As can be seen from *Figure 7.14*, the relationship between bound O_2 and pO_2 is not a simple linear one. The dissociation curve is sigmoidal, or s-shaped.

The simplest way of interpreting the oxygen–hemoglobin dissociation curve is to propose that there is some degree of cooperativity between the monomers which constitute the final molecule of hemoglobin – it should be remembered that there are four subunits to each molecule of hemoglobin. The binding of the first molecule of oxygen produces some conformational change in the structure of hemoglobin such that the binding of the second molecule of oxygen is aided, which in turn aids the binding of the third oxygen molecule. However, the binding of the fourth molecule of oxygen requires much larger amounts of oxygen to be present, due to the fact that fewer and fewer binding sites are available, which means that the probability of binding occurring is reduced. Hence, the dissociation curve reaches a plateau. The presence of the plateau region indicates that large changes in pO_2 have less effect on oxygen saturation at higher pO_2 values. For example, in *Figure 7.14*, increasing the pO_2 from 60 mmHg to 100 mmHg has little effect on the percentage saturation of hemoglobin with oxygen. In contrast, an increase from 20 mmHg to 60 mmHg has a dramatic effect.

One way of quantifying the affinity of O_2 for hemoglobin is to measure the P_{50}. This is the pO_2 which is required for 50% saturation of hemoglobin. The lower the P_{50} value, the greater the affinity of hemo-globin for O_2.

In addition to the cooperativity described above, there are a variety of other factors which influence O_2 binding to hemoglobin. Perhaps the most important factor is the pCO_2 (the CO_2 content) and H^+ content (i.e. pH) of blood. Essentially, these are the same thing, since CO_2 is an acidic gas. It dissolves in water according to the equation

$$CO_2 + H_2O \rightarrow H_2CO_3 \rightarrow H^+ + HCO_3^-$$

(H_2CO_3, carbonic acid; HCO_3^-, bicarbonate)

The effect of increased levels of CO_2 or H^+ (i.e. a reduction in pH) is to cause a shift of the O_2 dissociation curve to the right. This is termed the Bohr effect (*Figure 7.14*). This has important physiological consequences in that it causes the hemoglobin to give up any O_2 bound to it. This is particularly important in exercising tissues, for example, where there is an increase in CO_2 levels as a consequence of increased metabolism. The magnitude of the Bohr shift may be further enhanced if there is a degree of anaerobic metabolism, with the resultant production of lactic acid. The Bohr shift may also be elicited by a variety of organic phos-phate molecules, such as 2,3-bisphosphoglycerate (2,3-BPG). This com-pound is an intermediate in the glycolytic pathway. The levels of 2,3-BPG in humans living at altitude, where O_2 levels are reduced, are greater than in those living at sea level. This ensures that what O_2 is bound to hemoglobin is transferred to the tissues, thus facilitating the delivery of O_2 to the cells. In other animals, a variety of organic phos-phates have the same effect, such as ATP in fish, amphibians and some reptiles. Some fish show a greatly exaggerated shift in the dissociation curve, both to the right and downward, in response to increased levels of CO_2. This response is termed the Root effect.

When the magnitude of the Bohr effect is measured across a variety of mammalian species, it is found to be unequal, indicating that hemo-globins in different mammals behave in different ways. It has been found that the hemoglobin of small mammals (e.g. mice) is more sensi-tive to CO_2 and H^+ than the hemoglobin of, say, an elephant. Once again, this relates to the physiology of the animal. Mice have a larger surface area:volume ratio. The significance of this is that they lose their body heat at a faster rate. Given that they have an absolute require-ment to maintain a body temperature of 37°C, their metabolic rate, and thus O_2 consumption, must be increased to compensate for this loss, as the heat produced during metabolism maintains the constant body temperature. Therefore, the O_2 requirements per unit weight of tissue for a mouse are proportionately much higher than those of the elephant, which would have a much lower surface area:volume ratio and there-fore a reduced rate of heat loss. Therefore, a hemoglobin which is particularly sensitive to CO_2 and H^+ ensures that this demand for large amounts of O_2 in the tissues is met. Thus, the mouse has a very high metabolic rate. In contrast to this, the elephant has a much lower metabolic rate and O_2 requirement per unit weight of tissue and, as a consequence of this, its hemoglobin does not need to be so sensitive to changes in CO_2 and H^+.

7.6 Transport of carbon dioxide

The transport of CO_2 in blood is in some ways much simpler than the transport of O_2; for example, it does not require any specialized respiratory pigments to carry it. The reason for this is that CO_2 is far more soluble in solution than is O_2 – 20–30 times more soluble. However, like the transport of O_2, the transport of CO_2 occurs down concentration gradients and is a purely passive process.

The CO_2 that is produced in tissues enters the blood plasma and then passes into red blood cells. However, instead of simply dissolving in water, the CO_2 becomes bound within the bicarbonate ion (HCO_3^-). The reason for this is that when CO_2 dissolves in plasma, it undergoes the following reaction:

$$CO_2 + H_2O \rightarrow H_2CO_3 \rightarrow H^+ + HCO_3^-$$

Ordinarily, the formation of carbonic acid (H_2CO_3) is very slow. However, this reaction, which occurs inside red blood cells, is speeded up by the presence of an enzyme called carbonic anhydrase. The complete sequence of reactions is shown in *Figure 7.15*. A second way in which CO_2 is transported is in combination with hemoglobin. In this

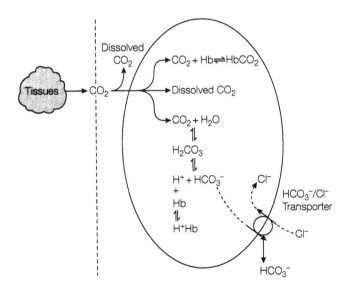

Figure 7.15. The fate of CO_2. At the lungs or gills, these reactions are reversed and CO_2 is liberated to the atmosphere.

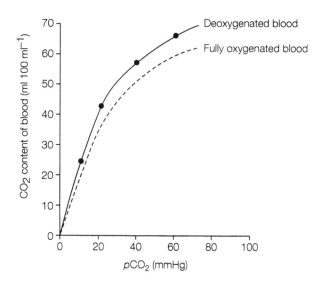

Figure 7.16. The CO_2 dissociation curve in deoxygenated and fully oxygenated vertebrate blood. Fully oxygenated blood has a reduced affinity for CO_2, therefore the curve moves to the right. This is known as the Haldane effect.

case, CO_2 combines with the free NH_2 group of the protein component. This results in the formation of carbamino compounds. In those animals without hemoglobin, other respiratory pigments (e.g. hemacyanin) function in an analogous manner.

In a similar fashion to O_2 transport, it is possible to construct a CO_2 dissociation curve *(Figure 7.16)*. It should be noted that the dissociation curve obtained in the presence of deoxygenated blood is shifted towards the left, so that deoxygenated blood has the ability to hold more CO_2 than oxygenated blood can. This shift is called the Haldane effect. The reason for this is that the hemoglobin of deoxygenated blood has more negative charges associated with it, and is therefore capable of buffering more H^+ *(Figure 7.15)*. This in turn promotes the conversion of CO_2 to HCO_3^-.

Further reading

Jones, J. D. (1972) *Comparative Physiology of Respiration*. Edward Arnold, London.

Levick, J. R. (1995) *An Introduction to Cardiovascular Physiology*. 2nd Edn. Butterworths, London.

Prosser, C. L. (1991) *Comparative Animal Physiology*. 4th Edn. Wiley-Liss, New York.

Gastrointestinal function

8.1 Introduction

The **gastrointestinal system** of an animal serves four functions. The first of these functions is feeding – that is, the delivery of food to the start of the gastrointestinal tract. This will be done in conjunction with other body systems, including the locomotory system and various components of the sensory system, such as the visual and auditory systems in animals which hunt and chase prey. The second function is digestion. **Digestion** is the process whereby the initial food that is eaten is broken down in both the physical and chemical sense of the word. The process of digestion permits the third aspect of gastrointestinal function to occur, which is absorption. The products of digestion are removed from the gastrointestinal tract by **absorption**, and ultimately transferred to all other cells of the animal, although this may not occur immediately. In many cases, the **nutrients** which are absorbed are held in storage until they are required by the animal. The final role of the gastrointestinal system is elimination or excretion. Foodstuffs which are eaten and cannot be digested are eliminated as waste products. This chapter will look at these aspects of the gastrointestinal system.

8.2 Feeding mechanisms

All animals are **heterotrophs** – they depend upon the ingestion of food to satisfy all their energy requirements. This contrasts with autotrophs (e.g. green plants) which are able to use the energy of solar radiation to convert inorganic molecules to organic molecules which are then metabolized. This is the process of photosynthesis. The foodstuffs which animals eat are tremendously varied, ranging from bacteria and planktonic life forms through to large animals. It is possible to make some broad generalizations about the feeding habits of animals and to therefore classify them into several groups (*Table 8.1*).

Table 8.1. Classification of animals by their feeding method

Nature of food ingested	Type of feeding method	Examples
Small particles	Pseudopodial – formation of digestive vacuoles	Protozoans, e.g. *Amoeba*
	Ciliate	Bivalve molluscs
	Tentacular	Some echinoderms, e.g. sea cucumbers
	Mucoid	Some urochordates
	Setous	Some crustaceans, baleen whales, flamingos
Large particles	Ingestion of inactive food	Some annelid worms, e.g. earthworms
	Scraping and boring mechanisms	Gastropod molluscs
	Seizure of prey	Carnivores
Fluid feeding	Sucking without penetration	Honeybees, hummingbirds
	Sucking with penetration	Leeches, vampire bats
Direct absorption of predigested foodstuffs	Absorption across the body surface	Endoparasites, some aquatic invertebrates
	Absorption from symbiotic partners	Corals, sponges, ruminant mammals

8.2.1 Ingestion of small particles

Ingestion of small particles, sometimes called **suspension feeding** is generally restricted to aquatic animals, the majority of which live in the marine environment. The reason for this is that there is a far greater abundance of potential foodstuffs, such as bacteria, algae, small invertebrates and so on, in the marine environment as opposed to the freshwater environment. It is a relatively common form of feeding, with examples in every phyla of animal which utilize this technique.

The simplest type of suspension feeding is phagocytosis used by the protozoans (*Figure 8.1*). This mechanism is also important in the bivalve molluscs and tadpoles (*Figure 8.2*). Here, particles are trapped as they pass through the gills and are directed towards the start of the digestive system by the action of cilia. At the mouth, the particles are sorted by size, allowing food to be distinguished from waste. The cilia also aid in the process of gas exchange by ensuring that water continually flows over the gills – without this there is the possibility that the water in contact with the gills could become depleted of oxygen. This illustrates the multi-functional role that gills – primarily respiratory structures – have. In other animals, as will be seen later, the gills are sometimes important in the mechanisms of osmotic and ionic regulation. In many cases, the process of suspension feeding is a continual process, as water has to pass continually over the gills in order for the

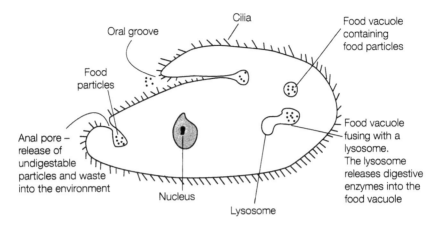

Figure 8.1. Feeding in a ciliated protozoan. Food particles enter a digestive vacuole where digestion takes place. Waste products are released into the environment.

animal to survive. Such continual feeding is sometimes seen in other suspension feeders, particularly those using mucus as a means of trapping food. Some **urochordates** produce mucus nets which trap food particles. So far, the examples provided have all been invertebrate or small vertebrate animals. However, suspension feeding is practised by some of the largest mammals, such as the baleen or right whales and some birds. In the case of baleen whales, rows of baleen plates are suspended from the roof of the mouth. When feeding, the mouth is filled with water and closed shut. Water is then squeezed out of the mouth and food is collected on the baleen plates (*Figure 8.3*).

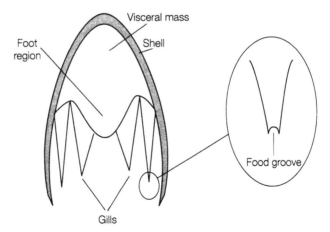

Figure 8.2. Cross section through a bivalve mollusc. At the tip of each gill are food grooves. Food particles are moved towards the food groove which then transports them to the mouth.

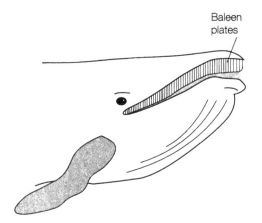

Baleen
plates

Figure 8.3. The baleen plates hang down from
the roof of the whale's mouth. As water is taken
into the mouth and expelled, food becomes
trapped in the plates and is then swallowed.

8.2.2 Ingestion of large particles

The simplest way to obtain large particulate food matter, is to eat inactive food masses. This has the advantage that the food does not need
to be caught. The most obvious material to eat is the environment in
which an animal lives. For example, the earthworm *Lumbricus* simply
eats the earth around it. The processes of digestion occur and nutrients are absorbed by the worm and waste material, feces, is excreted
from the animal, in the form of worm casts.

An alternative way of dealing with inactive food masses has been
developed in the gastropod molluscs (i.e. snails). In these animals, there
is a file-like organ called the **radula** which sits on a tongue-like structure called the **odontophore.** The radula moves to and fro, scraping off
food particles which are delivered to the beginning of the digestive
tract (*Figure 8.4*).

Animals that seize their prey are called **carnivores**. A carnivorous mode
of life is advantageous in that the food obtained is generally more
nutritious. This contrasts with a **herbivorous** mode of life, where it is
extremely difficult to digest the food which is ingested – remember
that plant cells have a cellulose cell wall enclosing all of their cells and
that cellulose cannot be digested without resorting to complicated
mechanisms. The disadvantage of a carnivorous lifestyle is that animals
must catch their food, which means that meals cannot be guaranteed.
Having obtained their food, they can either swallow it whole or chew
it up and swallow it in smaller pieces. The former method is seen in
snakes. These animals swallow food apparently larger than themselves
with the aid of modifications to the jaw. During seizure and swallowing
of the prey, the jaw dislocates and widens due to the fact that it is held
together with elastic tissue. In those animals that chew the prey before
swallowing there is an obvious need for teeth or teeth-like structures.
In cephalopods, for example, the mouth contains two beak-like struc-

tures that can reduce the size of food particles by a biting action. In some cases, capture of prey is aided by the presence of toxins in the saliva. The differentiation and specialization of tooth function occurs most obviously in mammals (*Figure 8.5*) and also in snakes. Snakes possess fangs which are specialized to deliver toxins which paralyze and kill captured prey. In some cases they may be hollow, rather like a hypodermic needle, or grooved to ensure delivery of toxin into the prey.

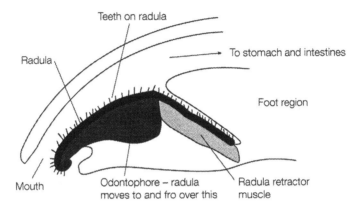

Figure 8.4. Feeding in snails. The movement of the radula scrapes particles of food off a larger food mass. The particles then pass into the digestive system.

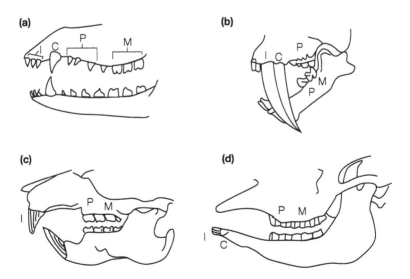

Figure 8.5. Mammalian teeth. (a) The general structure of mammalian teeth; (b) carnivore (sabre-toothed tiger); (c) rodent; (d) herbivore. I, incisor; C, canine; P, premolar; M, molar. Redrawn from Withers, P., *Comparative Animal Physiology*, 1992, Saunders College Publishing.

8.2.3 Fluid feeding

The easiest way to obtain fluids for feeding is to remove fluid from a source without having to penetrate the source concerned. This arrangement is typical of bees and hummingbirds which feed on nectar produced by plants. Even this apparently simple means of obtaining food may require specialist feeding apparatus. For example, the beaks of humming birds need to be elongated to enable them to reach into flowers to access the nectar. In other animals as diverse as leeches and vampire bats, the fluid may only be obtained by the piercing of some structure by specialized mouthparts. Good examples of piercing mouthparts are seen in insects, such as mosquitoes (*Figure 8.6*). Some insects have taken this approach to feeding even further – when they bite their prey, they inject digestive enzymes into it, and literally suck out the contents.

8.2.4 Direct absorption of nutrients

Perhaps the simplest and most convenient way of obtaining food is by the absorption of nutrients across the general body surface area. Such a mechanism generally means that the food absorbed is already digested or partially digested. **Endoparasites** that live in the guts of other animals, for example, absorb digested food from their host's digestive tract. Consequently, they have lost their own guts and their mechanisms for food capture as they no longer require them. Some free-living protozoans also obtain nutrients in this manner.

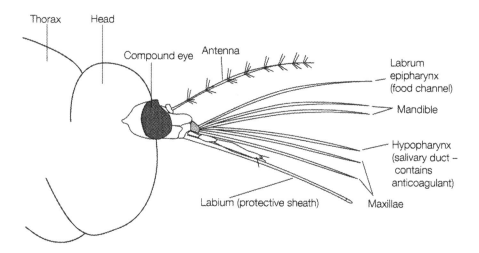

Figure 8.6. Feeding in the mosquito. The mandibular and maxillary stylets pierce the skin of the host. Anticoagulant is injected via the hypopharynx and the blood is taken up via the epipharynx. From Atkins, M.D., *Insects in Perspective*, 1978, Macmillan Publishing Co.

Some animals enter into a **symbiotic** partnership with other organisms to ensure that their food requirements are met. For example, some corals, sea anemones and bivalve molluscs live in symbiotic association with either green algae (Zoochloellae) or brown algae (Zooxanthellae). The amount of food obtained by the animal in this relationship varies. In some situations, the amount of energy captured during algal photosynthesis equals that consumed by the animal during respiration. At the other extreme are those relationships where the algae may produce only 5% of the animal's requirements. Another example of a symbiotic relationship which is essential for the delivery of nutrients to an animal is that of microorganisms and the digestion of cellulose. This will be dealt with later, but, essentially, animals are unable to digest cellulose since they lack the appropriate enzyme, cellulase. Digestion of this material is essential in herbivores, so, in order to overcome this problem, the guts of many herbivores contain symbiotic micoorganisms which are able to produce the cellulase.

8.3 The need for a gastrointestinal system

Having obtained food, an animal must digest it so that the nutrients can be utilized. Before gastrointestinal systems evolved, animals relied upon internal digestion. This is best exemplified by the protozoans. The process of digestion occurs within the cell in a **food vacuole**. Such a digestive process is also seen in some simple multicellular animals (e.g. the sponges). In one sense, internal digestion is beneficial since it means that the optimal environment for the process of digestion to occur can be easily maintained – for example, it is possible to maintain an optimum pH for enzymes in the vacuole. However, it should be remembered that this must represent an 'average' optimum pH since the many different enzymes required for digestion may have different optimum values. Internal digestion also has certain disadvantages. The first of these is that all digestive processes (i.e. protein, carbohydrate and lipid digestion) must occur within the same vacuole. This means that there cannot be any specialization of the digestive process. If this was the case, the animal would require separate vacuoles for protein digestion, carbohydrate digestion, and so on. Internal digestion also means that the digestive processes cannot be spatially or temporally separated – they must all occur simultaneously. The final problem with intracellular digestion is that it imposes a limit on the size of food particles that can be captured. Obviously, a protozoan would be unable to ingest a particle which is bigger than itself.

To overcome the problems associated with intracellular digestion, the gastrointestinal system and the process of extracellular digestion evolved. The gastrointestinal system can be thought of as a tube, the opening of which is the mouth, that passes through the centre of an animal and ends at a second opening, the anus. With extracellular digestion, the enzymes involved in the digestive process are secreted into the gastrointestinal system. The process of digestion occurs and the end-products of digestion are then absorbed back into the tissues of the animal. The advantage of the anus is that it ensures the one-way flow of materials through the gastrointestinal system, and allows the sequential processing of food that passes through. It also allows specialization of the gastrointestinal system, in the sense that some regions may be specialized for storage, some for digestion, others for absorption, and so on. This arrangement would be difficult to achieve if there was only one opening to the gastrointestinal system which would need to be used for both the intake of food and the removal of waste and indigestible products.

8.4 Generalized structure and function of gastrointestinal systems

The structures contained in a typical gastrointestinal system are shown in *Figure 8.7*. A typical gastrointestinal system consists of four basic regions.

(i)   a reception region;
(ii) a storage region;
(ii) a digestive and absorptive region;
(iv) an excretive and water absorptive region.

It is important to remember that not all gastrointestinal systems will contain all of the structures described above. The structure of the gastrointestinal systems of a variety of animals, showing their similarities and differences, is shown in *Figure 8.8*.

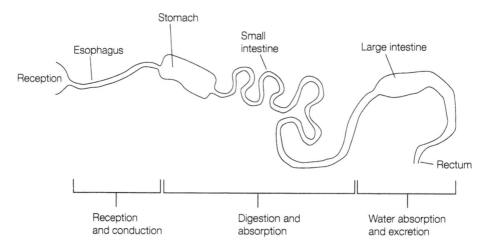

Figure 8.7. The basic components of a digestive system. Not shown are accessory structures, such as glands, which produce secretions that enter the digestive system and contribute to the process of digestion.

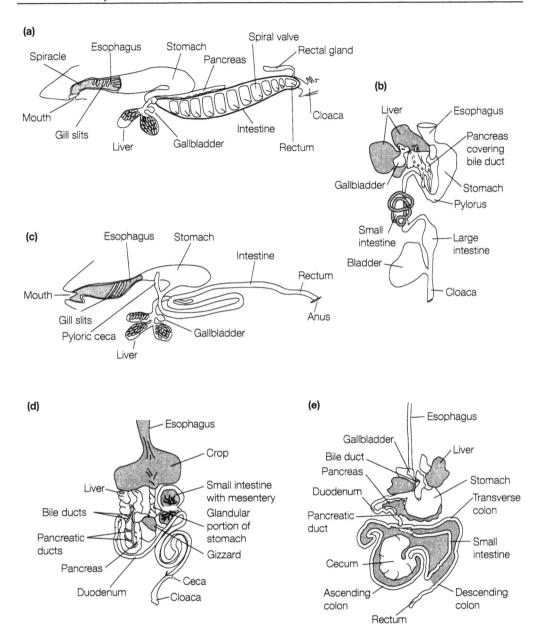

Figure 8.8. The digestive systems of various vertebrates. (a) Elasmobranch; (b) frog; (c) teleost fish; (d) bird; (e) mammal. Redrawn from Withers, P., *Comparative Animal Physiology*, 1992, Saunders College Publishing.

8.4.1 The reception region

The reception region of the gastrointestinal system is the mouth. In some respects, the process of digestion may be considered to begin here. In addition to structures involved in the gathering of food, such as teeth, which may have an important role in the mechanical breakdown of food, the mouth also secretes saliva. **Saliva** is an aqueous solution of mucin, a mucopolysaccharide. The primary function of this secretion is lubrication. In mammals, for example, it binds food particles together before they are swallowed and allows them to be swallowed more easily. Contained within saliva are a variety of other substances, which may include amylase (an enzyme which begins the process of carbohydrate digestion in some mammals), toxins (in the saliva of venomous snakes) and **anticoagulants** (in the saliva of blood sucking insects). The mouth and the remainder of the gastrointestinal system in virtually all animals, with the exception of the arthropods, are lined with mucus. This acts as a barrier and protects against both chemical and physical damage as the food particles pass through and are attacked by the various digestive enzymes. In the arthropods, the gut is largely lined with chitin which serves a similar protective function.

The **esophagus** is the tube which connects the mouth to the beginning of the gastrointestinal system proper. It may be considered as part of the receptive region, since it has no digestive function. The contraction of the esophagus is responsible for the movement of food particles down to the storage regions of the gastrointestinal system. In some animals (e.g. worms), it is the contraction of general body muscles, rather than the esophagus itself, which accounts for the movement of food. In most animals, though, the movement of food along the gastrointestinal track is achieved by the alternate contraction and relaxation of the circular and longitudinal smooth muscle which surrounds it. This process is known as **peristalsis**.

8.4.2 The storage region

The gizzard and stomach are enlargements of the anterior region of the gastrointestinal system and are, primarily, involved in the storage of partially digested food, although a small amount of digestion occurs here. The gizzard is a muscular sac which is involved in the mechanical breakdown of ingested food. It is seen in both invertebrates and vertebrates. In invertebrates (e.g. the arthropods) the gizzard may crush and filter food on the basis of size. This ensures that large particles of food do not pass down the gastrointestinal system but instead are retained in the gizzard for further mechanical reduction in size. In vertebrates with gizzards, mainly birds, the organ assumes the same role. Its contractile activity and subsequent mechanical breakdown of food particles is aided by the bird swallowing stones which help to crush particles of food.

The principal job of the stomach is to act as a store for partially digested food. The partially digested food at this stage is known as **chyme**. The stomach releases its contents, at appropriate intervals, into the remainder of the gastrointestinal system where the majority of enzymatic digestion occurs. In addition to its storage function, the stomach also plays a role in protein digestion by secreting enzymes called proteases which break down protein molecules. The environment of most vertebrate stomachs is highly acidic, with a pH of between 1 and 2. The high acidity of the stomach activates the **protease** enzymes which are stored and secreted as inactive precursors called zymogens.

In some herbivorous animals, e.g. cattle and sheep, the stomach has become highly specialized for the digestion of cellulose. In these animals, the stomach is multi-chambered, as opposed to the single-chambered stomach seen in other vertebrates. The structure of the stomach of these animals, known as ruminants, is shown in *Figure 8.9*. **Ruminants** are unable to digest the cellulose cell walls of the plant material they eat. The reason for this is that cellulose is a polymer of glucose, with individual glucose units joined to each other by β-glycosidic links. The enzyme system of vertebrates is only able to break α-glycosidic links. In order to digest the cellulose in their diet, ruminants have evolved a symbiotic relationship with bacteria and protozoans which live in the rumen and reticulum of the stomach and which are able to break β-glycosidic bonds. The microorganisms located in these chambers digest the cellulose that the animal initially eats. The partially digested food is then regurgitated for further chewing before being swallowed for a second time. This time, the food bypasses the rumen and reticulum (achieved by the presence of a fold in the rumen wall along which the food mass passes) and

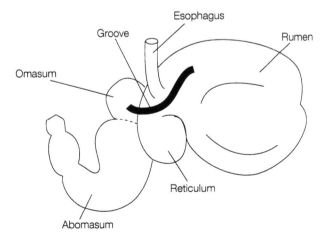

Figure 8.9. The four-chambered stomach of ruminants. The groove ensures that when food is swallowed for the second time, it passes through the digestive tract rather than remaining in the rumen.

enters the omasum and abomasum, which can be considered to be the 'true' stomach. The products of the breakdown of cellulose by micro-organisms results in the production of sugars, volatile fatty acids, methane and carbon dioxide, amongst others. The fatty acids and methane are waste products of the metabolism of microorganisms. However, the fatty acids are absorbed by the ruminant and utilized by them to produce energy. There are other nutritional advantages to this mode of life. For example, the microorganisms can produce protein from ammonia and urea, which will ultimately be digested and absorbed by the ruminant. Perhaps the biggest disadvantage is that although the cellulose is broken down by the symbiotic micro-organisms, much of it will be utilized by them in their own metabolic pathways. This means that the ruminant may not have ready access to sugars. This could have potentially catastrophic consequences because the brain requires a constant supply of glucose as it is unable to meta-bolize any other carbohydrates. In order to overcome this problem, the liver of ruminants has a high biosynthetic capacity for glucose – a process known as gluconeogenesis. The starting point in the synthesis of glucose in the liver is the fatty acids produced by the symbiotic microorganisms.

This type of arrangement between host and symbiotic microorganisms for the digestion of cellulose is also seen in invertebrates. For example, some beetles and termites contain protozoans in their guts which are responsible for the digestion of cellulose in their diets.

8.4.3 Digestive and absorptive region

The processes of digestion and absorption are undertaken in the midgut region of the gastrointestinal system – this is the region which consti-tutes the first part of the intestine. By this stage, food has been broken down by mechanical (i.e. teeth and mixing) means and is known as chyme. The digestive processes which occur from this stage onwards are enzymatic, and the aim is to break the major components of food-stuffs (i.e. carbohydrates, proteins and lipids) down into their constituent components which can then be absorbed and utilized by the animal. (It should be remembered that there are a variety of other nutrients required by animals, e.g. minerals, vitamins and so on. A full description of the requirements and roles of these substances is beyond the scope of this book). Given that there are three major types of food-stuff – carbohydrates, proteins and lipids – it follows that there are three major groups of enzymes associated with their digestion: **carbo-hydrases**, **proteases** and **lipases**, respectively. The role of each of these groups of enzymes will be considered.

Carbohydrate digestion. The enzymes responsible for the digestion of carbohydrates are known as carbohydrases. Their job is to break the **glycosidic bond** which links individual monosaccharides, to form di-, tri-, and polysaccharides (*Figure 8.10*). In general, carbohydrases may be divided into polysaccharidases and oligosaccharidases. Polysaccharidases are enzymes which digest polysaccharides such as starch, glycogen and cellulose, although the digestion of cellulose is dealt with in a special manner in most animals (see Section 8.4.2). The enzyme responsible for digesting starch is amylase. It is found in both vertebrates and invertebrates. In vertebrates, it is secreted in salivary secretions and also by the pancreas. In invertebrates it is secreted in saliva and also from glandular tissue found in the midgut region. The products of its digestion are glucose and maltose, an **oligosaccharide** which is digested further to produce glucose. Oligosaccharidases are enzymes which break down di- or trisaccharides. In some vertebrates (e.g. mammals), these enzymes, which include sucrase, maltase, trehalase and lactase, are located in close proximity to the epithelial cells of the gastrointestinal system which are responsible for carbohydrate absorption. This means that the carbohydrates are being digested almost at the same time as they are being absorbed. This implies that the processes of digestion and absorption are not as clearly delineated as was once thought. Oligosaccharidases are also found in invertebrates. However, invertebrate oligosaccharidases are less specific (in terms of the reaction catalyzed) than their vertebrate equivalents. For example, the vertebrate enzyme sucrase will split the disaccharide sucrose into its two component monosaccharides; in

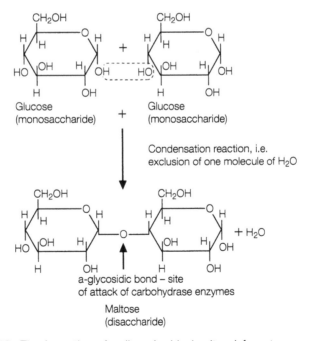

Figure 8.10. The formation of a disaccharide (maltose) from two monosaccharides (glucose). Some monosaccharides (e.g. cellulose) may be linked by an alternative glycosidic bond, the β-glycosidic bond. Animal carbohydrates are specific for α-glycosidic bonds, hence cellulose is indigestible.

contrast, invertebrate sucrase will break down both sucrose and maltose, into their component monosaccharides.

Protein digestion. The enzymes which are responsible for the break-down of proteins are called proteases. They break the peptide bonds which link individual amino acids together in a peptide or protein molecule (*Figure 8.11*). Protease enzymes are stored and sometimes released as inactive precursors called **zymogens**, which are activated on release. The reason for this is that the cells they are produced in, like any other cell, contain large amounts of protein. Therefore, the production of zymogens ensures that self-digestion does not occur. Proteases may be divided into two broad types. The first of these are the **endopeptidases**. These enzymes are responsible for breaking specific **peptide bonds** within a protein. The term specific, here, means that they are specific for the type of amino acid either side of the peptide bond. The other type of protease enzymes are the **exopep-tidases**, which are responsible for the removal of terminal amino acids – they remove either the N-terminal amino acid (aminopeptidases) or the C-terminal amino acid (carboxypeptidases).

It is possible to further subclassify the endopeptidases according to either the optimal pH at which the enzyme works or the nature of the amino acid that is required on either side of the peptide bond. There are a number of proteases which work at a low pH – the acid proteases – having optimum pHs of around 1.5–2.0. They include pepsin (stored and secreted as pepsinogen), which is released into the stomachs of all vertebrates, with the exception of the cyclostomes. The low pH acti-vates the zymogen, as can pepsin itself once it is activated. The ability of the activated form of an enzyme to activate the inactive zymogenic form is known as autocatalysis. In terms of its specificity, pepsin pref-erentially breaks peptide bonds between aromatic and dicarboxylic

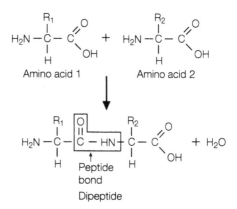

Figure 8.11. Two individual amino acids combine with the exclusion of water to form a dipeptide. The bond linking them together is called a peptide bond. The R group varies between individual amino acids, e.g. R_1 = H is glycine, R_2 = CH_2-CH-$(CH_3)_2$ is leucine.

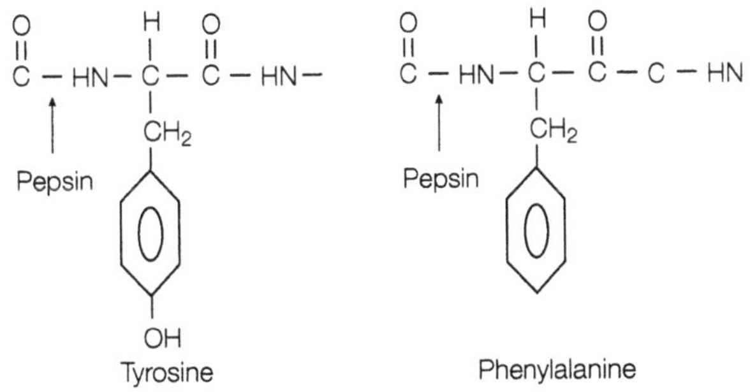

$$\begin{array}{ccccc} O & & H & O & \\ \| & & | & \| & \\ C & - HN - & C & - C & - HN - \\ \uparrow & & | & & \\ & & CH_2 & & \\ Pepsin & & & & \end{array}$$

OH
Tyrosine

$$\begin{array}{ccccccc} O & & H & O & & & \\ \| & & | & \| & & \\ C & - HN - & C & - C & - C & - HN \\ \uparrow & & | & & \\ & & CH_2 & & \\ Pepsin & & & & \end{array}$$

Phenylalanine

Figure 8.12. Pepsin is a specific endopeptidase. It attacks the peptide bond adjacent to the aromatic amino acids, e.g. tyrosine and phenylalanine.

amino acids (*Figure 8.12*). The presence of acid proteases in invertebrates is less well established. In addition to the acid proteases, there are a number of other proteases which have an alkaline optimum pH. Included in this group are trypsin and chymotrypsin. These are secreted as trypsinogen and chymotrypsinogen, respectively, by all vertebrates. Trypsin is activated by another gastrointestinal hormone called enterokinase, which is secreted by the gastrointestinal mucosa (the cells that line the intestine). In addition, trypsin is activated by autocatalysis – the conversion of some trypsinogen to trypsin promotes the conversion of further trypsinogen. Trypsin preferentially breaks the peptide bond involving arginine and lysine residues. There is some evidence for trypsin-like activity in the invertebrate phyla – for example, its activity has been demonstrated in some crustaceans and insects. Chymotrypsin is activated by trypsin. It breaks peptide bonds involving aromatic amino acids. Like trypsin, there is some evidence of chymotrypsin-like activity in the invertebrates, and again this has been demonstrated in some crustaceans and insects. In addition to these well established protease enzymes, there is evidence for protease activity specific to members of individual phyla. For example, unique proteases have been found in some crustaceans.

As indicated previously, there are known to be two types of exopeptidase (*Figure 8.13*). Aminopeptidases remove the N-terminal amino

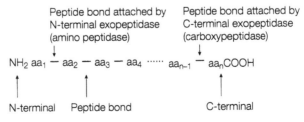

Figure 8.13. Exopeptidases remove one amino acid at a time (depending on whether they are specific for the C or N terminal), breaking a polypeptide or protein down into its individual amino acid components.

acid of a protein, whilst carboxypeptidases remove the C-terminal amino acid. The N-terminal is so called because the first amino acid in the chain has an amino group (NH_2) attached to it. The last amino acid has a carboxylic acid group attached to it and is therefore called the C-terminal. They are present in both vertebrates and invertebrates.

Lipid digestion. The digestion of lipids is achieved by the activity of two groups of enzymes: the **lipases** and the **esterases**. Lipases break down triglycerides, a common group of dietary lipids (*Figure 8.14*). In vertebrates, the major lipase activity comes from that secreted by the pancreas. In comparison, esterases are involved in the breakdown of structurally simpler lipids. The activity of these enzymes in mammals and in some invertebrates (e.g. crustaceans) is aided by the presence of **bile** which is secreted into the gastrointestinal system. Bile acts as a detergent, preventing fat droplets in the gastrointestinal system from aggregating together, thus increasing the surface area which is available for enzymatic attack. Other lipids utilized by animals include waxes, which are sometimes used as buoyancy aids in some marine inverte- brates. Although difficult to digest, some animals (e.g. fish and birds) which ultimately feed on these invertebrates, obtain a major proportion of their energy requirements from their digestion. In some animals (e.g. some moths), the digestion of lipids is achieved by the presence of symbiotic microorganisms in the gastrointestinal system. However, final digestion of the lipid is achieved with the use of enzymes produced and secreted by the animal itself.

Absorption. The products of digestion (amino acids, monosaccha- rides, free fatty acids and glycerol) must be absorbed from the gastrointestinal tract before they can be utilized by the animal. Much more is known about the processes of absorption in vertebrates than in invertebrates. The process of absorption requires the movement of these molecules from the lumen of the gastrointestinal tract, across the membrane and into the cells which make up the lining of the lumen. From here a variety of destinations await the molecules. The absorp- tion of substances from the gastrointestinal system may be achieved in one of two ways – it may be either a passive process or a carrier- mediated process.

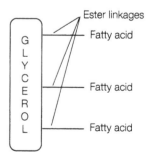

Figure 8.14. The basic structure of a triglyceride. It consists of a backbone of glycerol to which three molecules of fatty acids are linked via ester bonds. These bonds are the site of attack of lipid-digesting enzymes.

Passive processes, i.e. simple diffusion, rely upon the presence of a concentration gradient. This means that the concentration of the substance in the lumen of the gastrointestinal system must be higher than that of the interior of the absorptive cell. The process requires no energy expenditure, but will only continue until the concentrations of the substance in the lumen and cell are equal (i.e. an equilibrium has been established). Facilitated diffusion is also a transport mechanism that requires no energy expenditure and, like simple diffusion, materials will only be transported until there are equal concentrations of the substance on either side of a membrane. However, it does proceed to equilibrium much faster than simple diffusion. It relies upon a specific transport molecule, a protein embedded in the membrane, to transport the substance concerned across the membrane. In many ways, facilitated diffusion is a half-way house between simple diffusion and active transport. Active transport is another example of a carrier-mediated transport process; however, unlike the transport processes already described, active transport requires the input of energy in the form of ATP. Once again, it uses a membrane-bound transporter molecule which will only transport a specific substance, or, at most, a few structurally-related substances. Unlike the other two forms of transport, active transport does have the ability to transport substances against their concentration gradient. Remember that the other two transport processes come to a halt when there are equal concentrations of the substance on either side of the membrane. It is possible to distinguish between two types of active transport – primary active transport and secondary active transport. The two systems are summarized in *Figure 8.15*. Primary active transport mechanisms, the most common of

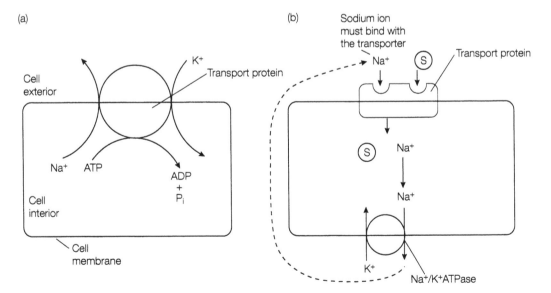

Figure 8.15. (a) Primary active transport via the Na+/K+ATPase pump. Three Na+ ions are pumped out of the cell in exchange for the entry of two K+ ions. (b) Secondary active transport. The active transport of substance S is dependent on the active transport of Na+ out of the cell.

which is the Na⁺/K⁺ ATPase pump, use a transport protein to move ions across the cell membrane against their concentration gradients. In the case of the Na⁺/K⁺ ATPase pump, the transporter protein contains binding sites for both K⁺ and Na⁺. It pumps Na⁺ out of the cell and K⁺ into the cell. As with all active transport mechanisms, it requires ATP in order to operate. In secondary active transport, the transport protein contains binding sites for a particular substance and Na⁺. Once both have bound to the transporter, a conformational change occurs which enables both Na⁺ and the other substance to cross the cell membrane. The Na⁺/K⁺ ATPase pump then removes Na⁺ from the cell so that it is free to combine again with the transport protein. Thus, the active transport of the substance is dependent on the active transport of Na⁺. Amino acids and glucose are examples of substances which are transported in this way.

Structural adaptations of the gastrointestinal system for absorption. The biggest adaptation to the absorption of nutrients shown by the gastrointestinal system is an increase in surface area, which ensures that absorption is maximized. This general adaptive feature is seen throughout the animal kingdom and takes several forms. For example, the gastrointestinal systems of many insects have a number of blind-ending sacs in the midgut region called **gastric ceca**, the function of which is to increase the surface area for absorption (*Figure 8.16*). The region of the gastrointestinal system in mammals where the majority of absorption occurs is the small intestine; in order to increase the surface area available for absorption, this region is highly folded. Originating from these folds are many finger-like projections into the lumen, again a functional adaptation which serves to increase the surface area available for absorption (*Figure 8.17b*). Other vertebrates have evolved alternative solutions to the requirement for a large surface area for absorption. For example, the elasmobranchs (the cartilaginous fishes, e.g. sharks and rays) have a so-called spiral valve which runs the entire length of the gastrointestinal system (*Figure 8.8a*). The

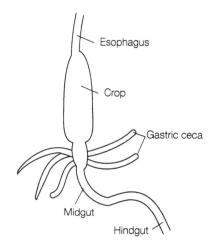

Figure 8.16. The gastric ceca of the insect digestive system.

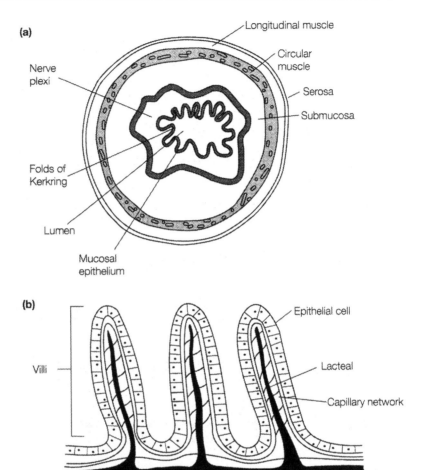

Figure 8.17. (a) The generalized structure of the mammalian small intestine. (b) The appearance of the mucosa of the mammalian small intestine. On the cells which make up the epithelium are even smaller finger-like projections called microvilli.

presence of this structure increases the area available for absorption, whilst at the same time hindering the movement of food, which means that the products of digestion are in contact with the absorptive regions for much longer.

The absorption of carbohydrates. The absorption of many sugars (e.g. glucose and galactose) occurs by the process of secondary active transport, whereby their transport is linked to that of Na^+ (*Figure 8.18*). This contrasts with the transport of other sugars, such as fructose, which is transported by the process of facilitated diffusion. Having entered the cells that constitute the absorptive region of the gastro-intestinal system, the monosaccharides must enter the general circulation of the animal. In mammals at least, the transport of

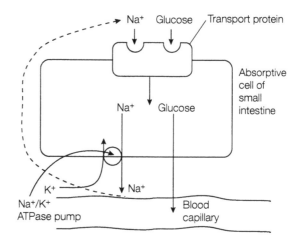

Figure 8.18. Glucose enters the cell by secondary active transport in conjunction with Na⁺. The glucose within the cell leaves via diffusion or facilitated diffusion and enters the circulation.

monosaccharides out of the cell that initially absorbed them occurs by either simple or facilitated diffusion.

The absorption of amino acids. The products of protein digestion, the amino acids, are also absorbed by secondary active transport in a process similar to that utilized in the absorption of monosaccharides. However, in some insects (e.g. cockroaches), the amino acids may be cotransported with K^+. The process of amino acid absorption is slightly more complicated than that of monosaccharides in that there are different transport mechanisms for different types of amino acid. Different transporters exist for the following amino acids:

● neutral amino acids;
● basic amino acids;
● acidic amino acids; and
● proline and hydroxyproline.

There is also evidence that some di- and tripeptides (i.e. molecules containing two or three amino acids, respectively) may be absorbed by a carrier-mediated transporter. Presumably, such a system must be fairly nonspecific. Given the 20 or so amino acids utilized by animals, it is possible to make several hundred unique tripeptide molecules, and it is unlikely that there is a specific transporter for each. Furthermore, there is evidence that whole proteins may be absorbed. This occurs by the process of **endocytosis** – the cell engulfs the protein molecule, internalizes it and digests it. This is thought to be the mechanism by which lactating mammals transfer immunity to their offspring. In this case, however, the proteins that are absorbed by the offspring are not digested, but instead enter the circulation of the offspring where they contribute to its immune defence system.

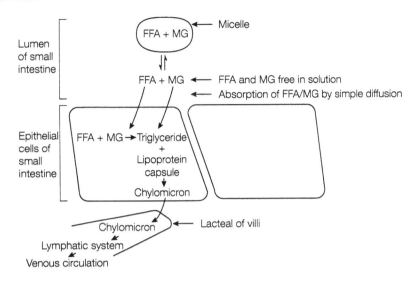

Figure 8.19. Absorption of the products of lipid digestion from the small intestine. The products of digestion are reformed before entering the lymphatic system prior to passing into the venous circulation. MG, monoglyceride; FFA, free fatty acids.

The absorption of lipids. The absorption of the products of lipid digestion (i.e. free fatty acids and monoglycerides) occurs by simple diffusion. The products of lipid digestion aggregate together with bile salts to form small particles called micelles. The quantity of monoglycerides and free fatty acids in 'solution' in the lumen of the gastrointestinal system is limited because of the nature of the products of digestion – they are hydrophobic molecules and will not freely dissolve in the aqueous solution that is contained within the lumen of the gastrointestinal system. However, those that are free in solution readily pass across the membranes of the cells which make up the lining of the gastrointestinal track – the epithelial cells in the case of vertebrates. Once these have been absorbed, further monoglycerides and free fatty acids are released from the micelles and are absorbed, and so the process continues. Thus, the micelles act as a 'store' of monoglycerides and free fatty acids waiting to be absorbed. Once absorbed, the free fatty acids and monoglycerides are converted back into triglycerides. The newly formed triglycerides are coated with a lipoprotein to form structures called **chylomicrons**. The chylomicrons enter the **lacteals** of the villi, which are a series of vessels that return excess tissue fluid to the systemic circulation via the lymphatic system. This process is summarized in *Figure 8.19.*

8.5 Excretion and water absorption

The final region of the gastrointestinal system is the hindgut. In vertebrates this comprises the large intestine. By the time food has reached here, the vast majority of digestion and absorption of nutrients has occurred. The material that enters the hindgut is usually semisolid – as opposed to the liquid nature of the rest of the contents of the gastro-intestinal system – due to absorption of water from the partially digested food (chyme) as it passes along the gastrointestinal tract. Some 80% of water is absorbed before it enters the hindgut, where the remainder is reabsorbed. This is facilitated by the active transport of sodium out of the gut lumen, accompanied by the passive movement of chloride ions, in order to maintain electrical neutrality. This generates an osmotic gradient between the lumen of the hindgut and the cells which line it, so water is absorbed by the process of osmosis. Osmotic mechanisms are described more fully in Chapter 9. The hindgut is normally colonized by a variety of bacteria. These bacteria produce nutrients, e.g. vitamins, as a consequence of their own metabolism which are absorbed by the animal. The remaining semisolid mass of undigested food is called feces, and this is expelled, periodically, into the environment.

Further reading

Morton, J. (1979) *Guts: The Form and Function of the Digestive Tract.* Edward Arnold, London.

Prosser, C. L. (1991) *Comparative Animal Physiology,* 4th Edn. Wiley-Liss, New York.

Sanford, P. A. (1992) *Digestive System Physiology,* 2nd Edn. Edward Arnold, London.

Osmoregulation

9.1 Introduction

By far the largest single component of any animal is water. It accounts for between 60% and 95% of the weight of an animal. The water within animals is located in various compartments – it may be inside cells in intracellular fluid (ICF) or it may be outside cells in the extracellular fluid (ECF). The ECF itself may be distributed between several smaller compartments, such as blood plasma and cerebrospinal fluid. Dissolved in these fluids are a variety of solutes, including ions and nutrients. It is of vital importance to animals that they maintain appropriate and correct amounts of water and solutes in their various fluid compartments. The ability to regulate water and solute concentrations is referred to as **osmoregulation**. It is intimately linked to the functioning of excretory organs, since this is one means by which water and solutes may be lost from the animal. It is the aim of this chapter to look at the osmoregulatory functions of animals, whilst Chapter 10 will consider excretory functions. Before doing so it is necessary to review the concept of **osmosis** as this is central to the process of osmoregulation.

9.2 The principles of osmosis

Osmosis is the movement of water across a selectively permeable membrane which separates two solutions, from a region of high concentration (i.e. a dilute solution) to a region of lower concentration (i.e. a concentrated solution). This process will continue until an equilibrium is established, at which point there is no further net movement of water and the concentrations of solution on either side of the selectively permeable membrane are equal. A selectively permeable membrane is one which allows only water to pass through it and no other substances, for example, solutes dissolved in the water. This is illustrated in *Figure 9.1*. In *Figure 9.1*, a solution of 1 M NaCl is separated from a 2 M NaCl solution (i.e. one solution is twice as concentrated as the other). In this case water will move from the 1 M solution (dilute) to the 2 M solution

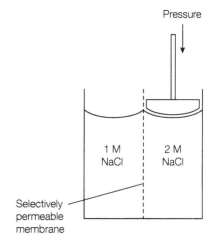

Figure 9.1. In this situation, water will move from left to right until concentrations are equal on both sides. Osmosis is the diffusion of water from a region of high water concentration (dilute solution) to a region of low water concentration (concentrated solution).

(concentrated) until the two solutions are of equal concentration. Osmosis is an example of a colligative property. A colligative property is one which is dependent upon the number of particles (i.e. ions, molecules etc.) dissolved in a solvent, and not, for example, upon their chemical nature. Obviously, this relates to the concentrations of particular solutions.

In *Figure 9.1*, if pressure is applied to the 2 M compartment, then the flow of water into it can be abolished – the pressure simply prevents water flowing from the 1 M compartment into the 2 M compartment. The pressure which is required to prevent the flow of water into a solution is called the **osmotic pressure** of the solution. In this case, the osmotic pressure of the 2 M solution is the pressure which must be applied to prevent any water movement into it. The osmotic pressure, then, is the pressure which prevents water movement, rather than being the cause of water movement. The concept of osmotic pressure can be rather confusing, therefore, rather than referring to the osmotic pressure which a solution exerts, it is more common to talk about the osmotic concentration of a solution. The more concentrated a solution is, the greater its osmotic concentration. It also follows that as osmotic concentration increases so does osmotic pressure. This is because the greater the concentration of a solution, the more water will flow into it and, therefore, the greater the pressure that will be required to abolish this flow.

In the example shown in *Figure 9.1*, the 2 M solution has a greater osmotic concentration than the 1 M and is said to be **hyperosmotic** to it. Equally, the 1 M solution has a reduced osmotic concentration in comparison to the 2 M solution and is therefore said to be **hypoosmotic** to it. When two solutions have the same osmotic concentration they are said to be **isosmotic**. The terms hyper-, hypo- and isosmotic say nothing about the composition of solutions. For example, a solution of 1 M KCl is isosmotic with a 1 M solution of NaCl for the simple reason that they both have the same number of particles (ions) dissolved in the solute.

The osmotic concentration of a solution must not be confused with the **tonicity** of a solution. Tonicity refers to the responses of cells when placed in differing solutions (*Figure 9.2*). When animal cells are placed in distilled water, the cell rapidly gains water by osmosis and will eventually burst. Distilled water is therefore **hypotonic** to the solution that is contained within the cell. Equally, if cells are placed in a concentrated salt solution, they will rapidly lose water by osmosis and will shrink. Thus, a concentrated salt solution is said to be **hypertonic** to the solution that is contained within cells. If a cell is placed in a solution and it neither gains nor loses water, the solution which surrounds the cell is said to be **isotonic** with the cell's contents.

9.3 Generalized osmotic responses of animals

It is possible to classify the osmotic responses of animals into two broad categories – they are either **osmoconformers** or **osmoregulators**.

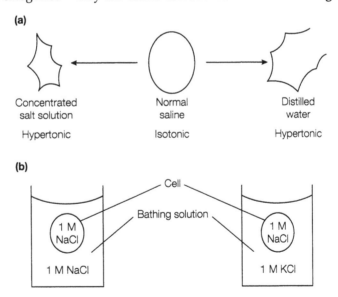

Figure 9.2. Tonicity. (a) When placed in distilled water, cells gain water and burst. When placed in a concentrated saline solution, cells lose water and shrink. (b) Isosmotic and isotonic are not the same. Left: the cell contents and bathing solution are isosmotic and isotonic so there is no net gain or loss of water. Right: the cell contents and bathing solution are isosmotic but not isotonic. If the cell membrane is selectively permeable to K^+, ions would enter the cell making its contents hyperosmotic to the bathing solution. Water would enter and the cell would burst.

Osmoconformers are animals whose body fluid concentration is exactly the same as that of the immediate environment in which they live. Typical osmoconformers include marine invertebrates, who have a body fluid concentration the same as that of salt water. This means that the two solutions (body fluid/sea water) are isosmotic. It should be remembered that although the solutions are isosmotic they do not necessarily have to have the same composition – although these animals may be in osmotic equilibrium, they do not necessarily have to be in ionic equilibrium and, therefore, they may have to expend considerable energy in maintaining the correct composition of body fluids (i.e. there is a requirement for ionic regulation). This will be considered later. The implication for animals that osmoconform is that if the osmotic concentration of the external environment changes, the osmotic concentration of the body fluids must change in a similar manner. This gives rise to a subgroup of osmoconformers which are able to tolerate wide changes in the osmotic concentration of their immediate environment, and are referred to as being **euryhaline**. The remainder are animals which can only tolerate much smaller changes in the osmotic concentration of their immediate environment, and they are referred to as being **stenohaline**.

The other major osmotic group of animals are those which are termed osmoregulators. These are animals which maintain a body fluid concentration that is different from that of their immediate environment. If the osmotic concentration of body fluids is maintained at a concentration greater than that of the immediate environment they are said to be hyperosmotic regulators (e.g. crabs); if they maintain their body-fluid concentration below that of the immediate environment they are said to be hypoosmotic regulators (e.g. some crustaceans). All terrestrial animals, by the very fact that they live on land, are osmo-

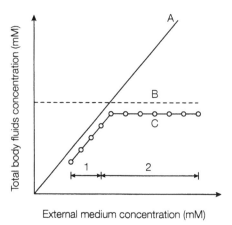

Figure 9.3. Organism A is an osmoconformer, organism B is an osmoregulator. Organism C is both an osmoconformer and an osmoregulator. Over the external concentration range indicated by 1 it is an osmoconformer, whilst over range 2 it is an osmoregulator.

regulators. The terms euryhaline and stenohaline apply to osmoregulators as well as to osmoconformers. The distinction between osmoconformers and osmoregulators is shown in *Figure 9.3*.

9.4 The osmotic responses of animals

It is convenient to consider the osmotic (and ionic) responses of animals in relation to the environments in which they live.

9.4.1 Osmotic regulation in marine environments

The vast majority of marine invertebrates are osmoconformers – the osmotic concentration of their body fluids is the same as that of the seawater they live in. This means that they are in osmotic equilibrium (i.e. there is no net gain or loss of water). However, as mentioned previously, this does not necessarily mean that they are in ionic equilibrium. Differences (even slight differences) in ionic composition between seawater and body fluids will result in the formation of concentration gradients. The resultant loss or gain of ions may challenge the physiology of the animal concerned and may also challenge the osmotic equilibrium. For example, an animal may gain ions from the seawater if a particular ion is at a greater concentration in seawater than it is inside the animal. This will result in the body fluids becoming hyperosmotic in relation to seawater and this in turn will result in the osmotic gain of water. The compositions of the body fluids of a number of marine invertebrates in relation to that of seawater are shown in *Table 9.1*.

In general, the osmotic concentrations of ions are not significantly different from the corresponding concentration in seawater. However, there are some exceptions, such as SO_4^{2-} and Ca^{2+}, which in some species may be present in concentrations markedly different from that found in seawater. This means that the concentrations of such ions need to be physiologically regulated – ions must be actively secreted or absorbed. In some marine invertebrates, for example, jellyfish, SO_4^{2-} ions are excreted to reduce the density of the animal, and, in doing so, the

Table 9.1. The ionic composition of marine invertebrates and seawater

Species						
	Na^+	K^+	Mg^{2+}	Ca^{2+}	SO_4^{2-}	Cl^-
Seawater	479.0	10.2	55.0	10.3	29.9	540.0
Jellyfish	464.0	10.6	54.0	9.8	15.5	567.0
Shore crab	500.0	12.0	30.0	24.0	16.6	550.0
Mussel	490.0	12.8	54.0	12.5	29.5	563.0

Ion (mmol l^{-1})

buoyancy of the animal increases. This occurs due to the fact that SO_4^{2-} is a relatively heavy ion. Thus, eliminating it from the animal reduces its weight and increases its buoyancy. Unregulated gains and losses may also occur, for example, across the general body surface area and the gills, through the ingestion of food and the production of waste substances (e.g. urine). Some invertebrates, for example, the octopus maintain body fluid concentrations that are hyperosmotic (i.e. more concentrated) to seawater, whilst others have body fluids that are hypoosmotic (i.e. less concentrated) to seawater, for example, the brine shrimp and a few other crustaceans. These examples are the exception rather than the rule when it comes to osmoregulation in marine invertebrates.

In contrast to invertebrates, the osmoregulatory options of vertebrates are different. It is possible to divide vertebrates into two major groupings – osmotic and ionic conformers or osmotic and ionic regulators. Examples of vertebrates who are in osmotic and ionic equilibrium with seawater include the hagfish. Hagfish belong to the cyclostomes and represent the most primitive vertebrates. In this sense, they behave in the same way as marine invertebrates. The osmotic and ionic conformation that the hagfish utilizes has been used as physiological evidence that vertebrates evolved in the marine environment. The majority of other marine fish, however, show varying degrees of osmotic and ionic regulation. The osmotic concentration of their **plasma** is approximately one-third that of seawater, therefore they are hypoosmotic regulators. The **elasmobranchs** (the cartilaginous fishes) have evolved a novel way of achieving this regulation. Given that their plasma is only one-third as concentrated as the seawater in which they live, they face two problems – the loss of water and the gain of ions. The loss of water is minimized by the animals achieving osmotic equilibrium by the addition of solutes to their plasma. The solutes added are **urea** and **trimethylamine oxide** (TMAO) (*Figure 9.4*). Urea is produced as an end-product of protein metabolism (see Chapter 10), whilst the biosynthesis of TMAO is less clear. In many cases, more urea and TMAO is added to the plasma than is necessary to produce osmotic equilibrium, thus making the plasma hyperosmotic to seawater. The result of this is that the animal gains water, in particular across the surface area of the gills. It should be remembered that the characteristics of gills which make them suitable for gas exchange – large surface area, thin walled, well vascularized – also make them ideal sites for the gain and loss of water and ions. This is advantageous to elasmobranchs, since it means that this 'excess' water can be used for the production of urine and the removal of waste products, such as excess ions that diffuse into the animal, which occurs, primarily, across the gills. Water gain also means that the animals do not need to drink seawater as a means to overcome potential water loss, and in avoiding this they avoid ingesting large amounts of salt that is dissolved in seawater, which would serve to further exacerbate the problems of ionic regulation.

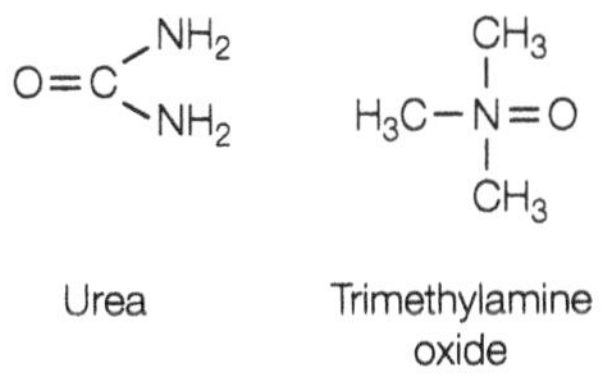

$$O=C \overset{NH_2}{\underset{NH_2}{\diagup}}$$

$$H_3C-\overset{\overset{\displaystyle CH_3}{|}}{\underset{\underset{\displaystyle CH_3}{|}}{N}}=O$$

Figure 9.4. The structures of urea and trimethylamine oxide (TMAO).

Urea   Trimethylamine
oxide

Potentially, the biggest problem with the addition of large amounts of urea to the plasma is that urea denatures and inactivates other plasma proteins. However, these animals have overcome this problem to such an extent that proteins and enzymes are unable to function correctly without urea. The second problem faced by the elasmobranchs is the gain of ions. Because their plasma has a different composition to seawater, a concentration gradient exist which favors the movement of ions into the animal – for example, there is a massive influx of Na^+ ions across the gills. Elasmobranchs overcome this problem with a special gland, the **rectal gland**, which is important in the excretion of excess Na^+. The rectal gland is a specialized gland which opens out into the rectum and secretes a fluid which is rich in NaCl. The small osmotic influx of water into these animals allows for the production of urine, which is another route by which excess NaCl may be excreted. A summary of the osmotic and ionic changes which occur in elasmobranchs is shown in *Figure 9.5.*

Marine **teleosts** (bony fish) face the same problems as elasmobranchs as their plasma is less concentrated than seawater. Loss of water, particularly across the gills, is compensated for by drinking large volumes of seawater. This solves one problem, but exacerbates another by adding a further salt load to the animal. This means that the animal must somehow excrete large amounts of NaCl. Since the kidney of teleost fishes is unable to produce a concentrated urine, there must be some other organ that is able to excrete large amounts of NaCl. This organ is the gill, which has a dual function in gas exchange and osmoregulation.

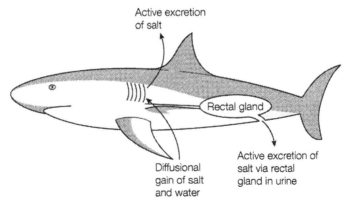

Active excretion
of salt

Rectal gland

Diffusional
gain of salt
and water

Active excretion of
salt via rectal
gland in urine

Figure 9.5. A summary of the major sites of loss and gain of water and salt in the shark.

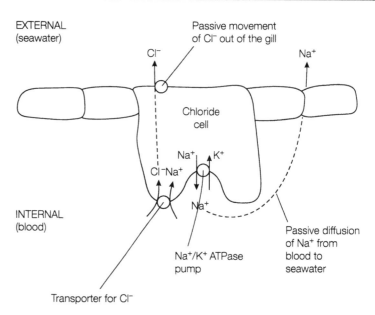

Figure 9.6. The mechanism by which NaCl is extruded from chloride cells in the fish gill.

The gills of these fish contain special cells, called **chloride cells**, which are responsible for the active transport of NaCl from plasma to seawater. The structure and function of chloride cells is shown in *Figure 9.6*. Essentially, Cl⁻ ions are actively extruded from the blood into the chloride cells, accompanied by the passive diffusion of Na⁺. From there, Cl⁻ moves passively out of the gill into the surrounding seawater. A diagram summarizing the osmoregulatory problems and solutions of marine teleost fishes is shown in *Figure 9.7*.

9.4.2 Osmotic responses in freshwater environments

The osmoregulatory problems faced by freshwater animals are the opposite of those faced by marine animals. Freshwater animals, by definition, must be hyperosmotic to the water in which they live. It would be impossible for any animal living in freshwater to be in osmotic and ionic equilibrium with it, unless the body fluids were made of distilled water. This means they face two problems – they tend to gain water from their immediate environment by osmosis and lose ions by diffusion due to the presence of large concentration gradients as only a minimal amount of solutes are dissolved in freshwater. Animals living in such an environment must be capable of significant osmotic and ionic regulation.

For both invertebrates and vertebrates, one way of limiting the gain of water (and loss of ions) would be to have an impermeable body surface. However, even if this were the case, water and ion movement across

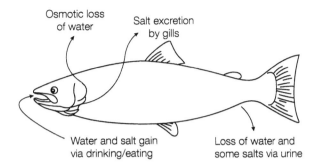

Osmotic loss of water

Salt excretion by gills

Water and salt gain via drinking/eating

Loss of water and some salts via urine

Figure 9.7. A summary of the osmotic and ionic problems of marine teleost fish.

gills could still occur relatively unhindered. The water that is gained by invertebrates is excreted as urine – the urine flow rate in freshwater invertebrates is much higher than that of corresponding marine species. However, the excretion of urine also results in the loss of ions and so exacerbates the diffusive ion losses which occur in these animals. In order to compensate for the loss of ions, active uptake mechanisms transport ions from the freshwater back into the animal. In many freshwater invertebrates, the site of ion uptake is not known and it is thought to occur across the general body surface area. However, in some invertebrates the site of uptake is known with some degree of certainty. In freshwater crustaceans, for example, it is known that active transport of ions occurs across the gills; in aquatic insect larvae, active transport of ions has been shown to occur in the anal gills.

Freshwater vertebrates face the same osmotic and ionic problems as freshwater invertebrates. When considering freshwater vertebrates, it is only necessary to consider the osmotic and ionic relations of the teleosts – there are very few elasmobranchs that are true freshwater species. Like invertebrates, the major site of osmotic water gain in teleosts are the gills. The excess water is removed by the production of large quantities of very dilute urine. Although the urine is dilute, it does contain some dissolved solutes, and because large volumes of urine are produced, urine excretion may result in a relatively large loss of ions. This in turn compromises the ion loss which is already occurring by diffusion from plasma to water. Some loss of ions can be compensated for by the gain of ions from food. However, the main source of ion gain is by the active transport of ions in the gills. It is thought that the transport of ions across the general body surface is insignificant. The osmotic and ionic relationships of freshwater teleosts is shown in *Figure 9.8*.

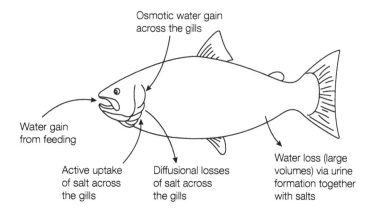

Osmotic water gain
across the gills

Water gain
from feeding

Active uptake   Diffusional losses   Water loss (large
of salt across   of salt across   volumes) via urine
the gills   the gills   formation together
                                                   with salts

Figure 9.8. A summary of the osmotic and ionic problems of freshwater
teleost fish.

9.4.3 Osmotic responses in terrestrial environments

With a few exceptions, the colonization of the terrestrial environment
has occurred in two groups of animals – the **arthropods** and the vertebrates. The ability to live on land has provided them with access to
increased amounts of oxygen, but poses a great threat to their water
and ionic balance. The reason for this is that on land there is limited
availability of water, so a major threat facing these animals is dehydration. Thus, life on land could be considered to be a compromise
between gas exchange and dehydration.

The cause of the greatest amount of water loss for terrestrial animals
is evaporation and such losses must be compensated for. A number of
factors influence evaporative water loss from an animal. These include:

- water content of the atmosphere – evaporation will be reduced as
  the water content (i.e. the relative humidity) of the atmosphere
  increases;
- temperature – evaporation will increase as temperature increases;
- movement of air over the evaporating surface – as air movement
  increases, so will the rate of evaporation;
- barometric pressure – as barometric pressure decreases the rate of
  evaporation increases;
- surface area – the larger the surface area exposed to the environment, the greater the water loss.

It should be noted that the relative importance of each of the factors
described above will vary for different animals. In addition, the factors
interact with each other.

For any animal, it is essential that, in the long term, it maintains a
balance between water loss and water gain. The possible routes for

Table 9.2. Possible routes for water loss and gain in terrestrial animals. The precise contribution of each of the components listed in the table makes to a particular animal or group of animals varies.

Water loss	Water gain
Evaporative loss (from the body surface and respiratory structures)	Drinking
Losses from urine	Water content of food
Losses from feces	Metabolic water
Loss from other secretions, e.g. saliva	Uptake across the general body surface

water loss and gain are shown in *Table 9.2*. Terrestrial animals have attempted to solve the osmotic problems of their way of life in a number of ways.

Terrestrial invertebrates. By far the largest proportion of terrestrial invertebrates are the arthropods – insects and spiders – and within these two groups, the insects are the most numerous. Other members of this phyla, the crustaceans, for example (with the exception of a few groups, such as woodlice), are predominantly aquatic animals.

One of the defining characteristic features of insects is the presence of an exoskeleton. The exoskeleton is covered by wax which forms the **cuticle** of the insect. The presence of the cuticle is one means by which evaporative water loss from the general body surface area can be reduced. However, it should be borne in mind that the cuticle is not totally impermeable to water, and there is still some water lost across it. Even so, the cuticle still represents a formidable barrier to evaporation. For example, evaporative water loss from the earthworm (which by comparison has only a very thin cuticle) is 70 times greater than the water loss from insects. Disruption to the arrangement of the waxes covering the exoskeleton, by physical or thermal damage, for example, results in increased water loss by evaporation. A second site of evaporative water loss from insects is the respiratory system, via the spiracles. Although many of the trachea which originate from the spiracles are covered with chitin, water loss from here may still represent a significant burden to the animal. In order to limit such loss, many insects utilize cyclic respiration which was discussed in Chapter 5. The loss of water via feces and urine production in insects is minimal. It will be seen later that urine produced by the Malpighian tubules enters the final regions of the gut, where absorption of water occurs before the urine and feces are discharged to the environment via a common orifice. This situation is aided further by the fact that insects excrete nitrogenous waste as **uric acid** – this is extremely insoluble in water, so it can be excreted with the loss of little water. A further adaptation to water conservation is seen in some insects (e.g. cockroaches), which, rather then excrete uric acid, deposit it in stores around the body, such as in the cuticle. This reduces even further the necessity to lose water when waste products are excreted.

The most obvious means of gaining water for insects is by drinking, for example, from rainfall, pools and so on. This is a source not available to all insects, such as those that live in hot arid environments and desert regions. Other potential sources of water include food, and the production of water during metabolism of food stuffs (**metabolic water**). In terms of water in food, perhaps the richest source of water is from plants – for example, the water content of fruits may be as high as 90%. When foodstuffs are fed into metabolic pathways to produce energy (i.e. ATP), water is produced as a by-product. The oxidative metabolism of 1 g of glucose produces 0.6 g of water, whilst 1 g of fat produces, on average, about 1 g of water. The final way in which some insects, such as cockroaches, are able to balance their water budget is to absorb water vapor from the air in the environment around them. In order for water to be absorbed in this way, the insect must have a very low body water content – 90% of its water must have been lost. In addition, the relative humidity of the surrounding air must be high – at least 80%. The actual mechanisms by which insects absorb water from the air and the site where it takes place are unclear.

Terrestrial vertebrates Terrestrial vertebrates comprise the reptiles, birds and mammals. It is possible to ignore the amphibians, who by their very nature are not truly terrestrial animals. Reptiles, which include snakes, lizards, crocodilians and tortoises, have dry, scaly skin which is well adapted to terrestrial life in that it represents a significant barrier to evaporative water loss. In addition, they excrete their nitrogenous waste as uric acid, which requires the loss of very little water. They are also able to produce very dry feces which further limits potential water loss. In terms of water gain, the drinking of water may present a problem because of the hot, arid environments where many of these animals are found. This means that water in food, together with water obtained during the metabolism of food, represents the most significant gain of water. Some lizards and tortoises produce a dilute urine which is stored in the bladder. This may be reabsorbed when the animals are dehydrated.

The adaptations seen in reptiles to maintaining a suitable water balance are also seen in birds. The water balance of birds may be further compromised by the fact that they must maintain a more or less constant body temperature. One way in which a heat-stressed bird may lose heat is to lose water by evaporation, which causes cooling. This may be achieved by the phenomenon of **gular fluttering** as described in Chapter 5. This is the rapid oscillatory movement of the mouth and throat which promotes water loss. It is analogous to panting in mammals. Gular fluttering represents a potential disruption to the animal's water balance, but, because birds are able to satisfy their water requirements by drinking for the simple reason that they are able to fly to find sources of water, this is not a significant problem. One potential problem with drinking water, particularly for marine birds, is that the water has a very high salt content. Together with the ingestion of large amounts of salt in their food, this means that these birds need to excrete large amounts of salt. In order to achieve this, marine birds have paired nasal **salt glands** (*Figure 9.9*). When the birds are facing a salt overload they begin to secrete a solution which is essentially a concentrated solution of NaCl. The glands are inactive until the bird becomes salt stressed. It should be noted that similar glands are found in reptiles, which, although primarily terrestrial animals, may spend part of their life in a marine environment – for example, marine iguanas and salt water crocodiles. Water loss by birds is further reduced by the fact that, like reptiles, they excrete a very dry urine (uric acid). This method of limiting water loss, which is so efficient that the water content of feces may be as a low as 25%, has undoubtedly contributed to the success of birds and reptiles living in hot, arid environments.

Mammals, like reptiles and birds described above, have the same potential routes for the loss and gain of water. The evaporative loss of water from the general body surface area is minimized by the presence of relatively impermeable skin and of fur and hair. Of greater importance, in terms of water loss, is evaporative loss from the respiratory tract – this may account for a large proportion of the water lost from an animal. However, mechanisms have evolved which serve to limit this loss. One such mechanism is breathing out air that is at a lower temperature than normal body temperature. This phenomenon is seen in all mammals. During inspiration, the walls of the nasal

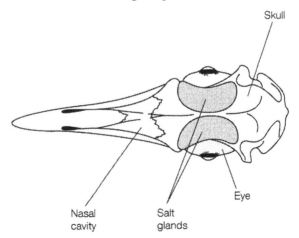

Figure 9.9. The salt glands in birds are located in the region of the eye socket and open into the nasal cavity. They secrete a solution which consists almost entirely of NaCl. Reprinted from Purves, W.K., Orians, G.H. and Heller, C., *Life: The Science of Biology*, 4th edn, 1995, p. 1011, with permission from Sinauer Associates.

passages transfer heat to the air entering the respiratory system. When the animal breathes out, warm air from the respiratory system passes over this cooled surface and condensation of water occurs. Water and salt loss also occur in those animals capable of sweating. In this situation, though, such losses are a means of regulating body temperature and not a true osmotic response. Water gain for many mammals is simply achieved by drinking. However, this is not possible for desert-dwelling mammals. The kangaroo rat (*Dipodomys spectabilis*), for example, does not drink, but survives on metabolic water – the oxidation of glucose, for instance, produces ATP, carbon dioxide and water.

Some mammals, for example, whales and dolphins, are exclusively marine-living mammals. It might be thought that such animals would face severe osmotic problems due to the gain of large amounts of salt from food. This may well be the case, but these animals have very efficient kidneys which can produce highly concentrated urine, thus ensuring that the excess salt they consume is excreted. However, it is not possible to produce a urine of infinite concentration. Generally, it is only possible to produce urine which is three- to four-times more concentrated than the plasma from which it has been formed. There is a very close link between osmoregulation and excretion, which is dealt with in Chapter 10.

Further reading

Lote, C. J. (1994) *Principles of Renal Physiology*, 3rd Edn. Chapman and Hall, London.
Rankin, J. C. and Davenport, J. (1981) *Animal Osmoregulation*. Blackie, Glasgow.

Excretory mechanisms

10.1 The need for excretory organs

The link between osmotic regulation and excretion was briefly made in the last chapter – animals which live in marine environments and consume large amounts of sodium chloride need to excrete excess quantities if they are to avoid disturbing their osmotic equilibrium. **Excretory organs** ensure that the composition of internal body fluids remain correct. By regulating the concentrations of solutes (Na^+, Cl^-, etc.) and water, osmoregulation is achieved. Excretory organs are involved in the removal of the waste products of metabolism, such as the breakdown products of nitrogen metabolism, and the removal of exogenous substances, such as drugs in humans. In short, the role of excretory organs is to balance the gains and losses of substances, so that if a particular substance is in excess in body fluids, its excretion is increased, and if its concentration is reduced, then its excretion is reduced.

10.2 Types of excretory organs

Virtually all animals have excretory organs. It is possible to classify excretory organs into two major groups: generalized excretory organs and specialized excretory organs. Generalized excretory organs include such structures as contractile vacuoles and various types of tubular structures, e.g. nephridial organs, Malpighian tubules and nephrons. Specialized excretory organs include structures such as salt glands (in gills and rectal glands – see Chapter 9), gills and the liver in vertebrates.

The physiology of the generalized excretory organs will be considered in this chapter.

10.2.1 Contractile vacuoles

Contractile vacuoles are the excretory organs of the coelenterates and the protozoans. The contractile vacuole cannot really be considered as an excretory organ in the sense that, say, the kidney is; rather it is an excretory organelle. Protozoans provide a good example of the mechanism of contractile vacuole function. They are present in all freshwater species of protozoan where there is a continual osmotic gain of water. However, their presence in marine species is less common.

Contractile vacuoles are spherical shaped organelles which water enters. The vacuole then fuses with the membrane of the protozoan and the water is expelled to the environment. As would be expected, the rate of water extrusion is related to the external osmotic concentration of the environment, and, therefore, the water influx. As the osmotic concentration of the external fluid surrounding the protozoan decreases (i.e. becomes more dilute), the rate of water entering the protozoan and, therefore, the amount of water that needs to be expelled increases.

The mechanism by which fluid enters the vacuole is not fully understood. However, it is thought that energy in the form of ATP is required for both the entry of water into the vacuole and for the emptying process. In the case of vacuole formation, active transport of water is thought unlikely. It is probable that ATP is required for the transport of ions across the vacuole membrane and it is this change in ionic concentration that results in the osmotic movement of water. Contractile vacuoles, or structures serving a similar function, are also found in the coelenterates. However, much less is known about the physiology of these structures compared to the contractile vacuoles in the protozoans.

10.2.2 Protonephridia

True excretory organs – discrete organs which have an excretory function – are found in all animals above the level of the coelenterates. The simplest of these excretory organs are found in the platyhelminthes and are called **protonephridia**. Protonephridia are excretory structures which exist as closed, or blind-ended, tubules and which do not connect with the **coelomic cavity**. The cell which forms the tip of the blind-ended tube is ciliated. If it contains a single cilia it is called a **solenocyte**, whereas if it contains several cilia it is called a **flame cell**. The organization of protonephridia in the platyhelminths is shown in *Figure 10.1*.

The precise way in which protonephridia work is still unclear. It is thought that the cilia within the tubule beat and create a negative pressure within the cell. This draws body fluid into the tip of the tubule, which comprises cells that contain a number of pores through which fluid flows. As the fluid passes through the cell membrane it is filtered.

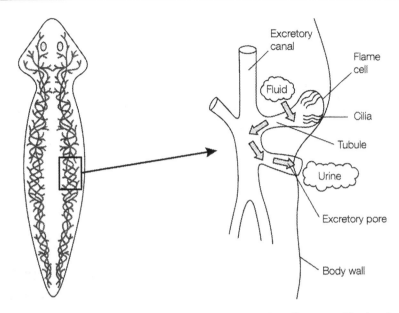

Figure 10.1. The arrangement of excretory organs in a flatworm. The beating of cilia draws fluid into the flame cell from the body cavity. As it moves along the tubule, its composition is modified before it is excreted.

Only small molecules are filtered through, whilst large molecules, such as proteins, are unable to cross the membrane. This process of filtration, which is termed **ultrafiltration**, is one of the basic processes which occurs in all excretory organs. The 'urine' which is finally formed in the protonephridia has a different osmotic concentration to that of the body fluids from which it was formed – it is usually more dilute than the body fluid from which it was formed. This indicates that the protonephridia are involved in the active reabsorption of substances from and secretion of substances into the fluid passing through them to produce the urine that excreted. Reabsorption and secretion alter the final composition of the urine making it quite different from that first formed by ultrafiltration. It should also be remembered that the active transport of ions into fluid which is destined to become urine will increase the osmolarity of that fluid. This results in urine which is more concentrated than body fluids, and water is likely to pass into this solution by osmosis. The important point to understand is that the active transport of ions can induce the passive movement of other substances, such as water. The processes of active reabsorption and secretion are two further basic aspects of excretory organ physiology which are common to all excretory organs.

10.2.3 Metanephridia

Metanephridia, sometimes called nephridia, are the organs of excretion found in many annelid worms. Metanephridia are defined as

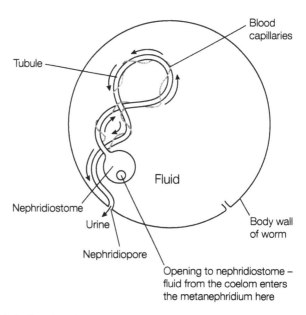

Tubule

Blood capillaries

Fluid

Nephridiostome

Urine

Body wall of worm

Nephridiopore

Opening to nephridiostome – fluid from the coelom enters the metanephridium here

Figure 10.2. Section through an annelid worm showing the arrangement of a metanephridium. There are two per body segment. Fluid enters via the nephridiostome and urine leaves via the nephridiopore.

excretory organs which have a ciliated opening to the coelom called the **nephridiostome** and which end in pores which open to the external environment, called **nephridiopores**. The structure of metanephridia, as seen in a typical annelid worm, is shown in *Figure 10.2*. Blood is filtered across the membranes of capillaries and the fluid produced enters the coelomic space. This is another example of ultrafiltration, since only water and molecules of small molecular weight enter the coelomic fluid, whilst larger molecules, such as proteins, remain in the vascular system. The fluid in the coelom then enters the metanephridia through the nephridiostome. The initial urine which is formed passes along the metanephridia where its composition is altered by the processes of reabsorption and secretion. The result is the production of a urine which is hypoosmotic (i.e. less concentrated) to the body fluids from which it was formed.

10.2.4 Malpighian tubules

Malpighian tubules are the excretory organs of the insects. The structure of a typical Malpighian tubule is shown in *Figure 10.3*. The precise number of these structures present in an insect will vary from just a few to many hundreds. As can be seen in *Figure 10.3*, Malpighian tubules have a closed end which lies in the fluid-filled cavity known as the **hemocoel**, and an open end which opens into the gut between the midgut and the rectum.

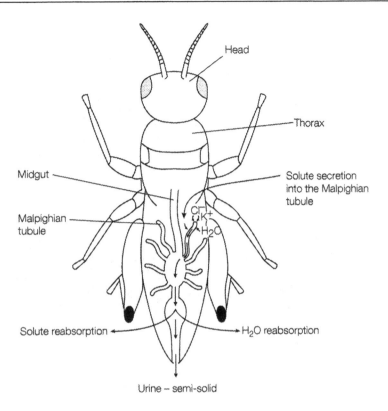

Head

Thorax

Midgut

Solute secretion
into the Malpighian
tubule

Malpighian
tubule

Cl⁻
K⁺
H₂O

Solute reabsorption

H₂O reabsorption

Urine – semi-solid

Figure 10.3. The arrangement of Malpighian tubules in a typical insect. K⁺ ions are secreted into the Malpighian tubule, which draws water and Cl⁻ ions in. The initial urine is modified as it passes along the tubule. In the rectum, large amounts of water are reabsorbed, producing very concentrated urine.

Because insects have an open circulatory system that operates at low pressure, there is no driving force for the ultrafiltration of body fluids. In this sense, the Malpighian tubule operates a little differently to all other excretory organs. In order to form urine, K^+ is actively transported from the hemocoel into the lumen of the Malpighian tubule. As a consequence, Cl^- ions follow, via diffusion and, ultimately, water due to osmotic potential differences. Other substances enter the tubule, including Na^+ ions, urates (nitrogenous waste products) and nutrients, such as amino acids. The urine then flows down the tubule and enters the midgut. It is possible that there is some modification of the composition of the urine as it passes down the tubule, but the vast majority of alteration occurs in the rectum prior to its discharge to the environment. In the rectum, both the nature and composition of the urine changes markedly. The concentrations of ions, such as K^+, are drastically reduced – in some cases by as much as 75%. More importantly, there is a tremendous amount of water reabsorption and the urates precipitate out as uric acid.

As will be seen later in this chapter, the fact that waste nitrogenous products are excreted as uric acid means that little water is lost in the process. What is thought to happen is that the active removal of substances from the urine into the cells of the insect is coupled to the flow of water, which follows ions into the cell by osmosis. In insects which are dehydrated, the amount of water that can be reabsorbed from the rectum is increased even further. Water reabsorption from the rectum is thought to be under the control of a neurosecretory hormone produced in the brain of the insect. It has been suggested that there are hormones which both increase and decrease water reabsorption in the rectum depending upon the degree of dehydration.

10.2.5 Crustacean green glands

Crustaceans (e.g. crabs and lobsters) have excretory organs called green, or **antennal glands**, located in the head region (*Figure 10.4*). The green gland consists of a blind-ending sac called the end sac connected to a tubule, the **nephridial canal**, which terminates in a region called the bladder. The bladder exits to the external environment via an excretory pore which is situated near to the base of the antenna. The end sac is surrounded by coelomic fluid which is filtered to produce the initial urine which lies within the gland. As with all excretory structures, the composition of the urine at this stage is similar to that of the body fluid (hemolymph) from which it was formed, with the exception that it contains no substances of high molecular weight, such as proteins. As this fluid passes along the nephridial canal, water and other solutes are reabsorbed. Unlike the bladder in mammals, for example, the bladder associated with the antennal gland is also capable of reabsorbing substances.

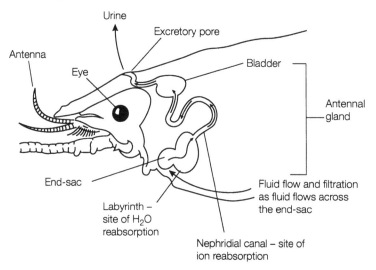

Figure 10.4. The head region of a crustacean and the location and arrangement of the antennal gland (green gland).

10.2.6 *The vertebrate nephron*

The principal excretory organ of the vertebrates is the kidney, the functional unit of which is the **nephron**. The typical mammalian kidney consists of about one million nephrons. With few exceptions (e.g. some teleost fish which have a secretory nephron), the nephrons of all vertebrates work on the principle of ultrafiltration followed by the active transport of substances into and out of the urine. Although sharing many common features, the nephrons of different species can serve different functions; for example, the nephron of desert living animals is highly adapted to conserve water, whilst that of freshwater fish has adapted to maximize water loss. The basic structure of the nephron is shown in *Figure 10.5*.

Nephron function. As indicated above, the majority of nephrons work on the principle of ultrafiltration, secretion and absorption. Filtration occurs in the **Malpighian body**, which is composed of a tuft of capillaries called the glomerulus together with the cupped end of the renal tubule, known as the Bowmans capsule. Overall, the pressure of blood within the capillaries is greater than the hydrostatic pressure of the fluid in the Bowmans capsule, and fluid is forced out of the blood and into the Bowmans capsule. This process is summarized in *Figure 10.6*. The

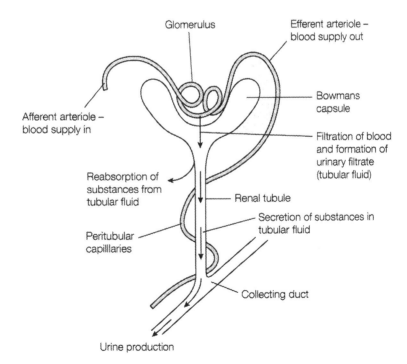

Figure 10.5. The basic components of the mammalian nephron. Filtration of blood occurs in the Malphigian body (the glomerulus and Bowmans capsule). The composition of the urinary filtrate produced here is altered as it passes along the renal tubule before being excreted as urine.

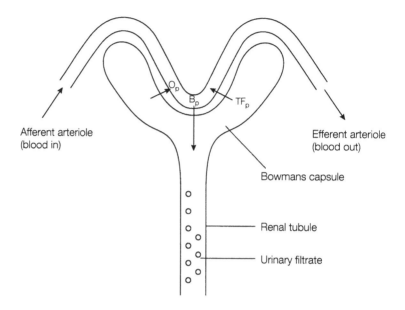

Figure 10.6. The fluid forces operating in the mammalian nephron. Overall, $B_p > O_p + TF_p$ so blood plasma moves from the glomerular capillaries into the Bowmans capsule and is filtered at the same time. O_p, colloid osmotic pressure of blood plasma; B_p, blood pressure in glomerular capillaries; TF_p, fluid pressure within Bowmans capsule.

initial urine which is formed is, essentially, protein-free plasma. As mentioned in Section 10.2.6, the nephrons of some fish do not use ultra-filtration to form the initial urine. Many such fish live in arctic waters, where one of the mechanisms that prevents the freezing of body fluids is the addition of antifreeze molecules to the blood, which lower the freezing point of blood and thus prevents it from freezing. This anti-freeze substance is a relatively small glycoprotein. If the nephrons of the fish worked on the principles of ultrafiltration, these molecules would potentially be lost in the urine when blood was filtered.

The renal tubule is the site of reabsorption and secretion of substances into and out of the urine. In fish, the renal tubule is divided into two regions: the proximal part and the distal part. Proximal and distal refer to the distance from the Malpighian body, proximal being nearest and distal being furthest away from the Malpighian body. The first part of the proximal region is the major site of absorption of solutes from the urine and the secretion of urea into the urine. Similar functions occur in the second part of the proximal region. The distal region, which is present in all freshwater fish, but only in some marine species, is the site of Na^+ reabsorption. This occurs by active processes but it may not result in the passive absorption of water. This results in freshwater fish producing large volumes (10-times larger then the volume produced by marine fish) of hypoosmotic urine. The urine of freshwater fish is 20-times more dilute than that produced by marine fish.

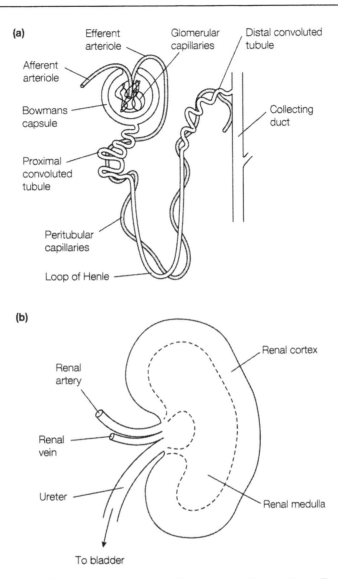

Figure 10.7. (a) The basic components of the mammalian nephron. The process of glomerular filtration occurs in the cortical region of the kidney. The loop of Henle dips down into the medulla. (b) The functional regions of the mammalian kidney.

In contrast to the nephron of freshwater fish, the nephrons of reptiles, birds and mammals conserve as much water as possible. In the case of birds and mammals, this enables them to produce urine which is much more concentrated than blood – mammals can produce urine which is up to 5-times more concentrated than blood. The structure of a typical mammalian nephron is shown in *Figure 10.7*. The reason why such a concentrated urine may be formed is due to the presence of the loop of Henle in the nephron. The **loop of Henle** works on the basis of a counter current multiplier system (*Figure 10.8*). NaCl is pumped

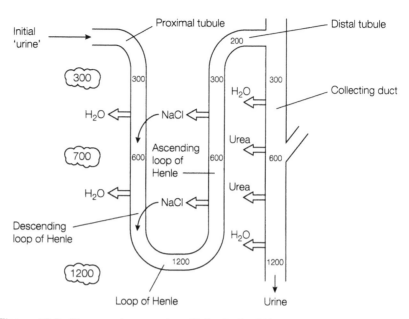

Figure 10.8. The countercurrent multiplier in the kidney.

out of the ascending loop of Henle by active transport. Water is unable to follow as the ascending loop is impermeable to water. NaCl moves from the tissue fluid into the descending loop of Henle by passive diffusion. As NaCl moves through the ascending loop it is once again pumped out, thus creating an NaCl concentration gradient. The filtrate that enters the collecting duct is less concentrated than when it first entered the tubule. Because the solute concentration of tissue fluids is greater then that of the filtrate, water leaves the collecting duct and the urine becomes more concentrated. The longer the loop of Henle, the greater the concentration gradient that can be generated, and the more concentrated the urine will be. There is a correlation between the length of the loop of Henle and the concentration of the urine produced – the greater the length of the loop the more concentrated the urine. Thus, mammals that live in deserts tend to have nephrons with very long loops of Henle which produce very concentrated urine. This reduces water loss. In mammals, water absorption is also regulated in the final part of the nephron, known as the collecting duct. Here, water reabsorption is under the control of ADH released from the posterior pituitary gland, which prevents urine production. Therefore, when this hormone is absent, mammals tend to produce large volumes of dilute urine. ADH works by altering the permeability to water of the walls of the collecting duct. In the presence of ADH, the water permeability of the collecting ducts is increased, and because the tissue surrounding the collecting duct has a high osmotic potential, water leaves the collecting duct. This results in the production of a small volume of highly concentrated urine.

10.3 Nitrogen excretion

The metabolism of carbohydrates and fats, two of the major nutritional components of animals, results in the production of carbon dioxide and water as waste products. Carbon dioxide is removed via exhalation from the lungs, as described in Chapter 5, and water is removed by excretory organs. In many cases, metabolism provides a valuable source of water, despite being considered a waste product.

In contrast, the metabolism of compounds containing nitrogen, primarily proteins and, to a lesser extent, nucleic acids, presents animals with more of a challenge. Animals are unable to store excess amino acids; therefore they must be metabolized. During metabolism, amino acids may be converted into substances which can subsequently be metabolized to produce glucose. This metabolism of amino acids, more correctly termed **deamination**, results in the production of ammonia. Deamination reactions may occur directly or they may be transdeamination reactions, whereby one amino acid is converted to another which is then deaminated to produce ammonia.

$$\underset{\text{Amino acid}}{R-\underset{\underset{NH_2}{|}}{\overset{\overset{H}{|}}{C}}-COOH} + H_2O \longrightarrow \underset{\text{Keto acid}}{R-\underset{}{\overset{\overset{O}{||}}{C}}-COOH} + NH_3$$

Similarly, when nucleic acids (purine and pyrimidines) are broken down, ammonia is ultimately produced as a consequence of their metabolism. Therefore, the animal has to deal with the newly-formed ammonia, which is a highly toxic compound. Animals deal with ammonia in one of three ways: by excreting it unchanged, by converting it to urea and excreting it, or by converting to uric acid and then excreting it.

10.3.1 Ammonia excretion

Animals which excrete ammonia unchanged are termed **ammonotelic**. Ammonia cannot be allowed to build up in an animal because it is extremely toxic – exposure of birds to levels as low as 10 ppm results in irritation of the respiratory tract, and body fluid levels as low as 5 mM are toxic to the majority of animals. It has several effects on animals. For example, it can alter intracellular pH and thereby affect intracellular metabolism as protein and enzyme function is altered, and it has also been suggested that it may alter mitochondrial function by interfering with the production of the proton gradients which are required for oxidative phosphorylation and the production of ATP.

Ammonia is extremely soluble in water and dissolves according to the following equation:

$$NH_3 + H_2O \rightleftharpoons NH_4^+ + OH^-$$

The excretion of nitrogen as ammonia is restricted to aquatic animals – since it is very soluble in water and has no difficulty crossing biological membranes. The fact that it is released into a large volume of water ensures that its toxic effects are minimized. Aquatic invertebrates can also excrete ammonia through their general body surface which is in contact with the water. However, in the case of fish, the majority of ammonia diffuses across the gills into the immediate environment.

10.3.2 Urea excretion

Urea is the main nitrogenous waste product excreted by terrestrial animals. Unlike aquatic animals, they cannot excrete ammonia as this requires large volumes of water into which it can be released. Urea has the advantage that although it is reasonably soluble, its toxicity is much reduced when compared with that of ammonia. Animals which excrete urea are said to be **ureotelic**.

Urea is synthesized via a series of reactions known collectively as the urea cycle (*Figure 10.9*). The production of urea is seen in a large variety of animals, both vertebrates and invertebrates. In many cases, the production of urea broadly follows the reactions shown in *Figure 10.9*,

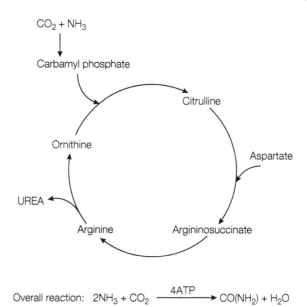

Overall reaction: $2NH_3 + CO_2 \xrightarrow{\text{4ATP}} CO(NH_2) + H_2O$

Figure 10.9. The Krebs cycle of urea production. In mammals, this occurs in the liver.

but the location of the enzymes involved may vary – in some animals they are located in mitochondria, whilst in others they are located in the cytoplasm. In some cases (e.g. in the sharks and rays), the urea that is produced is not excreted, but is reabsorbed and enters the circulatory system where it contributes to the total osmolarity of the blood and is essential for osmoregulation. Rather than being a waste product, urea is essential to these animals – without it they would not survive.

10.3.3 Uric acid excretion

The third major means by which nitrogenous waste products are excreted is in the form of uric acid. Animals which excrete uric acid are said to be **uricotelic**. Animals which are uricotelic include insects, birds and reptiles. The advantage of excreting uric acid is that very little water is lost with it – the urine of birds, for example, is a semi-solid white sludge. It may be considered to be an adaptive mechanism to living in terrestrial environments where the conservation of water may be at a premium. The metabolic pathway by which uric acid is formed is very complex. Essentially, ammonia is converted into a substance called glutamine, which in turn provides the structure around which uric acid may be synthesized. The structure of uric acid is shown in *Figure 10.10*.

10.3.4 Environmental patterns of nitrogen excretion

From the preceding text, it appears that the pattern of nitrogen excretion is closely related to the habitat of the animal. To a large extent this is true. Aquatic animals can simply release ammonia into their environment, thus escaping its high toxicity and the need for expensive (in terms of metabolic energy requirements) conversion to other substances. Terrestrial animals on the other hand, living in environments where water may be less readily available, excrete uric acid – thus excreting excess nitrogen whilst, at the same time, conserving water. However, in reality, animals usually produce more than one type of nitrogenous waste product. Humans, for example, produce both ammonia and uric acid in small amounts, although they are considered to be ureotelic. Different life stages of a single species may also

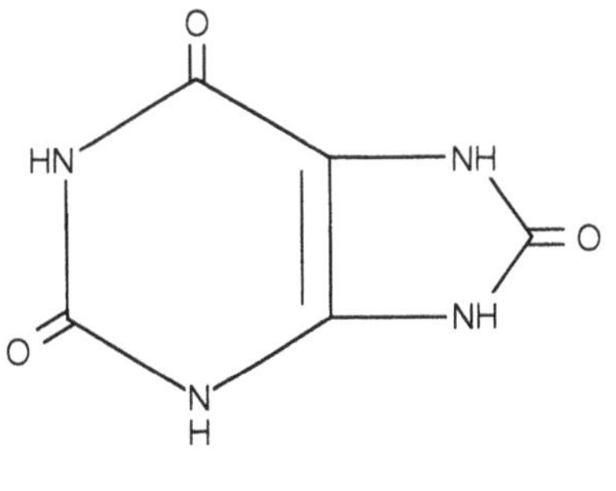

Figure 10.10. The structure of uric acid.

secrete different nitrogenous waste products. For example, juvenile amphibians (i.e. tadpoles), which are aquatic, excrete ammonia, whilst the adult form, which is semi-terrestrial, excretes urea.

Further reading

Prosser, C. L. (1991) *Comparative Animal Physiology*, 4th edn. Wiley-Liss, New York.

Reproduction

11.1 Introduction

Reproduction is one of the key defining characteristics of life. The ability of an animal to produce viable offspring is intimately linked to the evolutionary development of an animal. The rearrangement of **genes**, which may occur during reproduction, gives rise to variation which is the life blood of the evolutionary process. This also implies that reproductive processes are linked to genetics and development. This chapter will concentrate on aspects of reproductive physiology only, and the reader is directed elsewhere for further information regarding genetics, development and evolution. The two fundamental physiological processes of reproduction are **asexual reproduction** and **sexual reproduction**.

11.2 Asexual reproduction

Asexual reproduction requires only one parent animal, and does not involve the production and utilization of **gametes**. Gametes are the egg and sperm cells which will be discussed later. Since asexual reproduction only involves one parent, all the offspring produced as a result of this process are genetically identical to the parent. Thus, asexual reproduction has little part to play in the production of genetically diverse offspring and the evolutionary development of the animal concerned. In general, asexual reproduction is limited to the simpler invertebrate groups, e.g. protozoans, coelenterates etc. There are several forms of asexual reproduction, perhaps the simplest being **binary fission** (*Figure 11.1*). Binary fission is the reproductive process used by protozoa. Reproduction is achieved by **mitotic chromosome division**, whereby the chromosome number of the animal is doubled. The cell (since protozoans are unicellular animals) then divides into two, producing two daughter cells. Each daughter cell contains a full set of **chromosomes** and goes on to develop into a mature cell. This process may be further developed in some protozoans, such as *Plasmodium* sp., the causative agent of malaria. In this protozoan, rather than a single

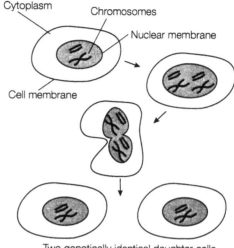

Figure 11.1. Binary fission in a single-celled organism. The number of chromosomes doubles and the cell divides into two daughter cells (mitotic cell division).

mitotic division of the parent cell and the production of two daughter cells, there are multiple mitotic divisions resulting in the production of many daughter cells. The production of many identical daughter cells is termed schizogany. Another example of asexual reproduction is vegetative growth, or budding. This is seen in many coelenterates, e.g. corals. In this situation, the offspring are produced as a result of the growth and development of the parent animal – an extension, or bud, is produced (*Figure 11.2*). This bud then develops into a new animal. This process is sometimes called **stolonization** and is a reproductive process common to plants.

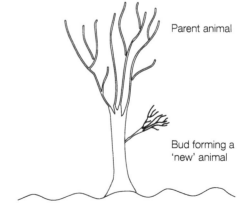

Figure 11.2. The process of budding. An entire new animal is formed as an outgrowth of the parent. Connection between the two is lost and an individual animal is formed.

Asexual reproduction could be regarded as an inferior form of reproduction as it results in little genetic variability and is restricted to simple invertebrates. However, it does have certain advantages. For example, it is not necessary to go to the biological expense of producing gametes or finding a mate. Furthermore, it is a very rapid way of reproducing – the doubling time of protozoans utilizing this method may be as little as 30 minutes.

11.3 Sexual reproduction

In contrast to asexual reproduction, sexual reproduction requires two parent animals. In the vast majority of animals there are separate sexes, this situation being known as the **dioecious** state. The parents must produce specialized sex cells called gametes, which must fuse together in the process of **fertilization** to produce a **zygote** which then ultimately develops into the adult animal. The gametes – eggs or ova in females and sperm in males – are formed in specialized sex organs called **gonads** and are haploid – they contain half the total number of chromosomes that non-gamete cells (e.g. muscle cells) possess. This means that at some stage in their production, the cells from which they have originated have undergone a **meiotic cell division**. This results in a halving of the chromosome number (*Figure 11.3*). It is this meiotic cell division which is so important in producing offspring which are not identical to their parents, since there is a degree of 'reshuffling' of

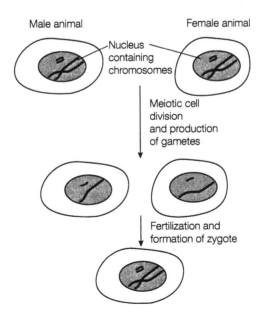

Figure 11.3. The production of gametes by meiotic cell division and the formation of a zygote.

the parental genes, sometimes called crossing-over, from which the offspring are produced. Hence, the offspring are not identical to their parents.

Some animals (e.g. snails) have both male and female gonads. Such animals are said to be **hermaphrodite**, and are **monoecious** in contrast to the separate sex dioecious animals mentioned earlier. Generally, hermaphrodite animals tend to avoid self-fertilization, preferring fertilization by another animal. A variation on this theme is **parthenogenesis**. This is the development of an egg without fertilization by sperm. Parthenogenesis is displayed in many animals, both invertebrates (e.g. arthropods) and vertebrates (e.g. some fish). In some instances of parthenogenic reproduction, the offspring produced are genetically identical to their parent, because the egg from which they have developed arose from mitotic cell division as opposed to meiotic cell division, where there is some reshuffling of the genes. Parthenogenesis of this type is seen in aphids, for example. During the summer months when environmental conditions are ideal for the animal, vast numbers of aphids will be produced by mitotic parthenogenesis. During this time only female aphids are produced.

Sexual reproduction is often regarded as the better way of producing offspring, mainly because the offspring that are produced are not genetically identical to their parents, so variation is achieved. This could be considered of value in an environment which is changing – some offspring will be better suited to the environment than others and will therefore survive and, in turn, reproduce themselves. However, it is costly in the sense that animals must invest in the production of gametes and must find a mate.

11.4 Gamete production

As indicated previously, sexual reproduction is dependent on the production of gametes. Gametes are produced from specialized cells called primordial germ cells. These cells are found in the gonads (i.e. ovaries and testes) and are destined to become the germ cells. In many animals, particularly the invertebrates, little is known about the formation of gametes. Much more is known about this process in vertebrates.

11.4.1 Sperm production

Spermatozoa (sperm) are the gametes produced in male animals. They are produced in the male gonads, the testes. By far the majority of testes, in both invertebrate and vertebrate animals, are tubular in nature. The structure of a typical mammalian testis is shown in *Figure 11.4*. Sperm are produced in the walls of the tubules and as they develop and mature, they migrate towards the lumen of the tubule.

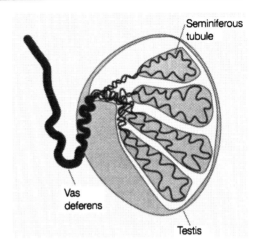

Figure 11.4. The gross organization of the mammalian testis (simplified).

They eventually enter the lumen in order to be released, either into a female or the immediate external environment. The mammalian testes are made up of hundreds of tubes called seminiferous tubules, which are the site of sperm production. A cross section through the wall of a seminiferous tubule shows that on the very periphery of the tubules are the undifferentiated and diploid **spermatogonia**. Differentiation of these cells takes place close to the lumen. Eventually, fully formed sperm cells are liberated into the lumen of the tubule (*Figure 11.5*). The final appearance of the sperm cells varies from animal to animal.

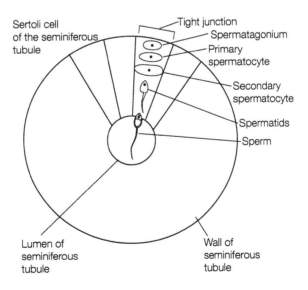

Figure 11.5. Section through a single seminiferous tubule showing the stages of sperm production. Spermatogonia and primary spermatocytes are diploid cells whilst the remaining cells are haploid. The phenomena of spermatogenesis occurs in the tight junction between adjacent Sertoli cells.

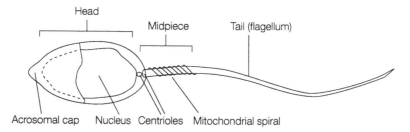

Figure 11.6. Organization of a sperm cell. The head region contains the nucleus (haploid) and the acrosome. The acrosome contains enzymes that enable the sperm cell to digest the membranous barriers which surround the egg. The midpiece contains mitochondria which provide the energy for motility.

However, with a very few exceptions (e.g. some invertebrates), the sperm cell consists of three basic regions: the head, the midpiece and the tail. The generalized appearance of a sperm cell is shown in *Figure 11.6.*

Much of what is known about the control of sperm production comes from studies on mammals. In the testes are a group of cells called **Sertoli cells**. These cells are the ones principally involved in the process of spermatogenesis, being involved in the nutrition of developing

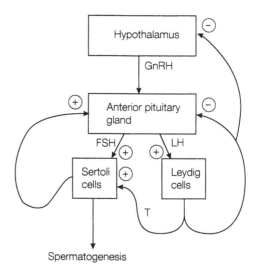

Figure 11.7. The hormonal control of sperm production. Sertoli and Leydig cells are found in the testes. Sertoli cells are principally involved in spermatogenesis, supplying nutrition to developing cells, removing dead cells and producing fluid. GnRH, gonadotropin-releasing hormone; FSH, follicle-stimulation hormone; LH, luteinizing hormone; T, testosterone; I, inhibin; +, stimulatory; –, inhibitory.

sperm cells and the removal of dead sperm cells. The Sertoli cells are stimulated by **follicle-stimulating hormone** (**FSH**) which is released from the anterior pituitary gland. FSH release is promoted by **gonadotropin-releasing hormone** (**GnRH**), which is released from the hypothalamus. FSH is thought to stimulate the Sertoli cells to release substances which promote spermatogenesis. The Sertoli cells are also stimulated by testosterone, which is produced by the Leydig cells of the testes. Testosterone is a steroid hormone and it is essential for the growth and development of the male reproductive system. The release of testosterone is under the control of another anterior pituitary hormone, called luteinizing hormone (LH). Like FSH, LH is itself under the control of GnRH. The control of sperm production in mammals is summarized in *Figure 11.7*.

11.4.2 Egg production

Egg or ova production is characteristic of biologically female animals. Eggs are produced in the female gonads – the ovaries. Generally, egg cells are much larger than sperm cells. This is primarily due to the accumulation of nutrients within the egg. The structure of the ovary may take several forms. For example, some worms and insects have ovaries which appear as tubular structures. In this case, the tip of the tube is the site of production of egg cells. In contrast to this are the ovaries of mammals (*Figure 11.8*). In this case, eggs develop in the **ovary** from where they are released. Subsequently, they are collected in the oviduct and pass along the rest of the reproductive tract. The reproductive tract of the human female is shown in *Figure 11.9*. What happens to the eggs after they have been produced depends upon the type of animal. The vast majority of invertebrates and some vertebrates (birds, reptiles, fish) simply deposit the eggs in their immediate external environment. Fertilization of the egg may occur before or after they have been released – it may be either external or internal. Animals which display this type of behavior are said to be **oviparous**. Alternatively, eggs may be retained in the body of the animal where fertilization occurs, and the fertilized egg stays in the animal whilst it develops. Whilst the embryo is developing, it obtains all its nutrients from the egg The young animal may either hatch from the egg inside the mother and be born live, or it may hatch soon after the egg is laid. This type of behavior is seen in some invertebrates, such as some annelid worms, and also in some vertebrates (e.g. sharks) and is termed **ovoviparity**. Another option is for the egg to be retained in the female animal and for fertilization to occur internally. However, rather than surviving on nutrient stores laid down in the egg, the embryonic animal forms a close anatomical relationship with its mother which is the source of all nutrients – in the case of humans, this is the placenta. This type of behavior is seen in mammals and is termed **viviparity**. There are advantages to an ovoviviparous or viviparous way of life in that the offspring are relatively well developed at birth.

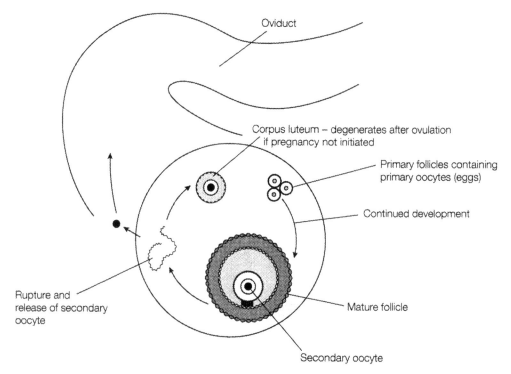

Figure 11.8. A summary of the events occurring during egg development in humans. The primary oocytes (diploid) are laid down during embryonic development. The first stage of meiosis occurs during development and is then inhibited to be completed just prior to ovulation. The final stage of meiosis is only fully completed once the egg cell is penetrated by a sperm cell.

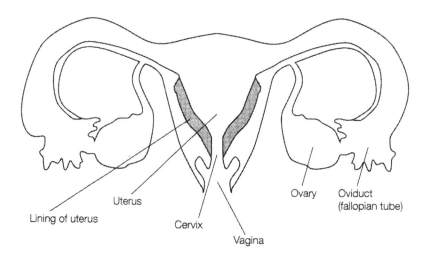

Figure 11.9. The simplified organization of the human female reproductive tract.

The disadvantage is that it is much more costly in biological terms for the parent, in the sense that a relatively high degree of parental care is needed after the birth of the young.

The control of egg production is much better understood in mammals. Generally, egg production in mammals is cyclical in nature. The **estrus cycle** is the cycle of egg production which is present in the vast majority of mammals. This contrasts with the **menstrual cycle**, which is observed only in primates. There are several differences between the two cycles. For example, in the menstrual cycle, females are fully receptive to males for mating. However, the egg which is released approximately half-way through the menstrual cycle, is only viable for fertilization for about 72–96 hours after its release. This contrasts with the estrus cycle, where the female is only receptive to the male for a brief period during the cycle. A second major difference is what happens to the lining of the reproductive tract at the end of the cycle. In the menstrual cycle, the lining disintegrates and is shed; in the estrus cycle this does not occur. The organization of the human menstrual cycle is shown in *Figure 11.10*.

11.5 Fertilization

Having produced gametes, the next step in sexual reproduction is fertilization – the union of gametes. The result of this union is a fertilized egg known as the zygote. As development of the zygote continues it becomes an embryo. The process of fertilization may occur before or after the release of eggs into the external environment – it may be either internal or external.

The process of external fertilization is somewhat risky. Large numbers of eggs must be released, and, consequently, large numbers of sperm to ensure that some of the eggs are fertilized. The eggs, both fertilized and unfertilized, are a potential food source to other animals and many of the potential offspring will not survive. In contrast to this, internal fertilization (and development) helps to ensure that some of the embryos remain viable and are born. In order to facilitate internal fertilization, sperm must be transferred from the male to the female. The organ that ensures that this is achieved is the penis. The penis exists in both invertebrates and vertebrates. For example, male insects transfer sperm to the female of the species and store sperm in a structure called the spermatophore. There is tremendous diversity both in the anatomy and physiology of the penis.

Much of our knowledge of the biochemical events of fertilization derive from studies of invertebrates, in particular the sea urchin. As usual, much of the knowledge gained from one animal has been extrapolated to other animals. For example, it is known that only one sperm will fertilize a single egg and fertilization by more than one sperm cell

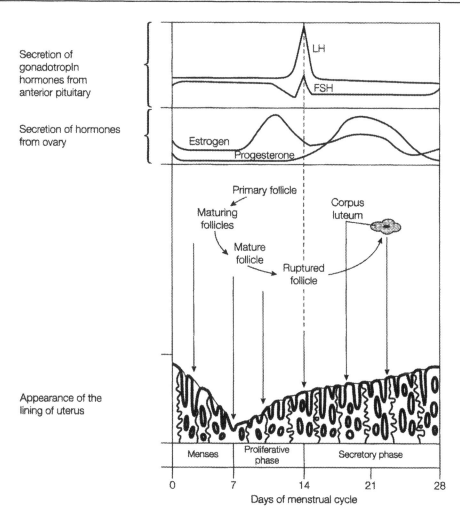

Secretion of gonadotropin hormones from anterior pituitary

Secretion of hormones from ovary

Appearance of the lining of uterus

Figure 11.10. Summary of the human menstrual cycle. Follicle development is induced by the release of FSH. This results in the secretion of estrogen by the developing follicle and promotes development of the lining of the uterus. LH is responsible for the process of ovulation. Under the influence of LH, the empty follicle (corpus luteum) secretes progesterone. If fertilization fails to occur, this secretion is inhibited and menstruation begins. From Hickman, Roberts and Larson, *Integrated Principles of Zoology* 10th edn, p. 643, 1996, Wm C. Brown Communications Inc. With permission of The McGraw-Hill Companies.

(polyspermy) is prevented. In the sea urchin, once a sperm has fertilized an egg, the egg produces a fertilization membrane. This results in all other sperm being removed from contact with the egg. Following fertilization, the nuclei of both the egg and the sperm fuse, and changes occur in the cytoplasm of the zygote, which starts to divide and grow to form the embryo. In some animals (e.g. the sea urchin) the first divisions of the zygote begin only a couple of hours after fertilization has occurred. In contrast, in mammals, the first division may occur 24–36 hours after fertilization.

11.6 Development and pregnancy

11.6.1 Initial cleavage of the zygote

The zygote undergoes the process of **cleavage**. Cleavage is the division of the nucleus and cytoplasm of the zygote which results in a mass of cells, several thousands in many cases, called **blastomeres** collected together in a spherical structure called the **morula**. Several factors influence the process of cleavage, one of which is the amount and distribution of yolk within the zygote. Yolk is a nutrient which allows the zygote to survive until either the egg has hatched, or the egg implants in the mother who then becomes the source of nutrients. Three types of eggs may be identified:

(i) Isolecithal eggs seen in, for example, marsupials, placental mammals, molluscs, echinoderms and other, particularly aquatic, invertebrates. The egg contains a small amount of yolk which is evenly distributed throughout the cytoplasm. The cleavage occurring in these eggs is referred to as being **holoblastic**. This means that the lines of cleavage pass throughout the entire cell. The physiological significance of isolecithal eggs is that they contain limited reserves of food, i.e. little yolk. Therefore, the developing zygote must find alternative sources. In the case of placental mammals, the mother provides food and this will be discussed in more detail later. In the case of aquatic invertebrates, the zygote develops into a free-living larval form which in turn metamorphoses into the adult form.

(ii) Telolecithal eggs, seen in birds, reptiles, fishes, molluscs and flatworms. The yolk of the egg is collected at one end called the vegetal pole. The other end of the egg, called the animal pole, contains the nucleus and cytoplasm. The cleavage pattern seen in these eggs is **meroblastic** – this means that the lines of cleavage are incomplete. In this case, they halt at the divide between yolk and cytoplasm for the simple reason they are unable to pass through the yolk.

(iii) Mesolecithal eggs. This type of egg is seen in amphibians. It contains an amount of yolk that is intermediate between the two types of egg described previously. Like isolecithal eggs, mesolecithal eggs undergo holoblastic cleavage. However, the yolk tends to aggregate at the vegetal pole and the lines of cleavage have difficulty passing through this region.

In addition to holoblastic and meroblastic cleavage, zygotes may undergo either radial or spiral cleavage. In the case of **radial cleavage**, the cell divisions which occur produce layers of cells one on top of another. Alternatively, cleavage may be said to be **spiral**, where groups of cells are produced that lie in the grooves between other cells rather than directly on top of each other. These differences in cleavage patterns are one means by which animals may be grouped together or classified. For example, spiral cleavage is observed in animals such as the annelids, some flatworms and some molluscs. These animals are referred to as **protostomes**. This means that the first 'opening' which forms in the animal – the **blastopore** – is the mouth. This contrasts with radial cleavage seen in the vertebrates and the echinoderms, for example. These animals are referred to as **deuterostomes**. In this case the blastopore forms the anus and the opening to the mouth occurs secondarily.

In some animals, for example, the urochordates, arthropods and molluscs, the fate of each of the cells of the blastula is predetermined from a very early stage. Thus, the removal of a single cell will result in the formation of a defective animal that will lack a particular structure, depending upon which cell was removed. This type of development is known as **mosaic development** – each blastomere will eventually form a particular part of the adult animal and all blastomeres are required. In contrast, echinoderms and vertebrates display what is known as **regulative development**. In this case, if the blastomeres are separated early enough in cleavage, each will continue to develop into (ultimately) the adult form.

11.6.2 Gastrulation – the formation of body layers

Gastrulation represents the movement and rearrangement of the blastomeres into locations and relationships that are appropriate for further development of the zygote. Cells on the exterior of the blastula (a mass of cells with a hollow center) move inwards, forming a structure known as a **gastrula**. Gastrulation represents the further movement and organization of cells within the developing animal. At the completion of gastrulation, the embryo (as it is now known) consists of three germ layers, each of which develops into specific structures in the fully-formed animal. The process of gastrulation is broadly the same in all animals; however, the precise mechanism by which it is achieved varies between different egg types, generally being more complex in telolecithal zygotes.

The gastrula is now referred to as an embryo rather than a zygote. It is seen to be composed of three layers of cells: the **endoderm**, the **ectoderm** and the **mesoderm** (*Figure 11.11*). These layers will give rise to specific structures in the fully formed animal. The endoderm will produce the epithelial lining of the gastrointestinal tract and respiratory system, the ectoderm will give rise to the skin and the nervous system, and the mesoderm will form the musculoskeletal system, the circulatory system and the excretory system. The formation of specific organ systems is called organogenesis. This requires the further development and differentiation of individual groups of cells.

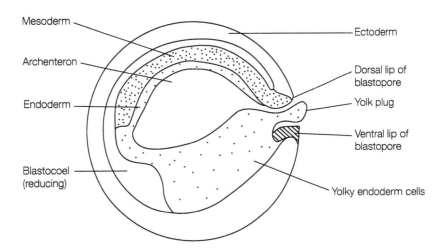

Figure 11.11. The frog gastrula.

11.6.3 Extra embryonic membranes and the placenta

The development of birds, reptiles and mammals is characterized by the presence of four extra-embryonic membranes, so called because they are not part of the embryo, although they are derived from it, and are discarded after (or, in some cases, before) the embryo is born. The extra-embryonic membranes are shown in *Figure 11.12*.

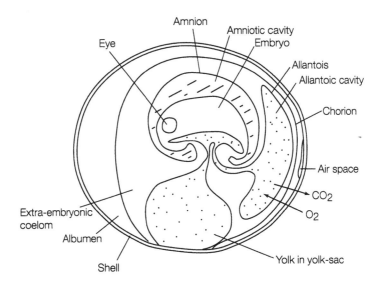

Figure 11.12. Extra-embryonic membranes in a bird egg.

The first of the extra-embryonic membranes is the **yolk sac**, a membrane which surrounds the yolk. In some cases, for example, in mammals, it is reabsorbed before the embryo is born. In others, such as fishes, it remains attached to the free-living fish larva and is used as a food source until the fish can begin to feed itself. The **amnion** is a membrane that surrounds a fluid-filled cavity in which the embryo develops. The third extra-embryonic membrane is the **allantois**. This is a structure which is derived from the hindgut of the developing embryo. Its function is to act as a storage site for some waste products, particularly nitrogenous waste products, and also acts as a site of gas exchange. The final extra-embryonic membrane is the **chorion**. This is the outer membrane and encircles all the other structures. In mammals, where fertilization and development is internal, the development of the embryo is dependent upon the presence of the **placenta**. This is a structure which is produced by the reorganization of the extra-embryonic membranes and is essentially the life-support system for the developing embryo.

Following fertilization in mammals, which occurs in the oviduct, the fertilized egg (zygote) moves down to the uterus (*Figure 11.9*). The blastula stage of the mammalian embryo is known as the **blastocyst**. It consists of an outer layer of cells called the **trophoblast**, and an inner group of cells, called the inner cell mass, from which the embryo and fetus will develop. A few days after fertilization, the blastocyst embeds itself in the lining of the uterus; this is a process known as implantation. This is aided by the trophoblast cells secreting proteolytic enzymes, which allow the blastocyst to bury itself into the **endometrium**, which becomes well vascularized. This enables the developing embryo to obtain its nutritive requirements from its mother. However, as the embryo continues to divide and grow, this arrangement becomes difficult to sustain. Therefore, a placenta develops, which is a mixture of both maternal and fetal tissues, and this becomes the life-support system for the developing animal. The outermost layer of the trophoblast develops into the chorion and is the embryonic component of the placenta, whilst the layer of endometrial cells immediately adjacent to the chorion is the maternal component. Contact between the embryo and mother is enhanced by the development of chorionic villi. These are finger-like projections of the chorion into the endometrium, which serve to increase the surface area available for the uptake of nutrients, gas exchange and so on. The fetal capillaries which are present in the chorionic villi originate from arteries in the embryo and eventually drain blood into veins which re-enter the embryo. These arteries and veins are collected together in the **umbilical cord**, which is formed in part from the allantois, and whose role is to connect the embryo to the placenta. The amniotic cavity is formed from a space between the trophoblast and the inner cell mass. This becomes fluid-filled, and thus affords protection to the developing embryo. The relationship between the embryo and placenta is illustrated in *Figure 11.13*.

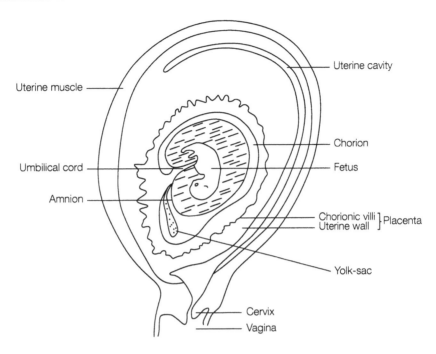

Figure 11.13. The mammalian fetus in the uterus.

One other significant role of the placenta of mammals is as an endocrine organ. The trophoblast in the placenta secretes the hormone chorionic gonadotropin. The role of **chorionic gonadotropin** is to ensure that the **corpus luteum** continues to secrete estrogen and progesterone. The corpus luteum represents the remains of the follicle from which the egg has been released during ovulation. The presence of estrogen and progesterone ensures that no further eggs and follicles are allowed to develop during the pregnancy.

11.7 Birth

Birth, or parturition, is best understood in mammals, although the full details of what initiates parturition are not fully understood. In mammals, birth is achieved via the coordinated contraction of uterine smooth muscle, together with other muscles, and the expulsion of the fetus through the vagina. This process is aided by another endocrine substance secreted by the placenta, called relaxin. **Relaxin** increases the flexibility of the bones of the pelvis and dilates the cervix (the birth canal).

What triggers the initiation of uterine contractions is not clear. What is known is that there are significant endocrine changes which occur in the placenta. The first of these is a decrease in the ratio between progesterone and estrogen levels in the maternal plasma. In some mammals, for example, sheep, this ratio is altered by the increased activity of the fetal adrenal gland. The secretion of cortisol from the fetal adrenal gland promotes the synthesis of estrogen at the expense of progesterone – both compounds being produced from a common precursor. The significance of this is two-fold. Firstly, progesterone is a potent inhibitor of uterine smooth muscle contraction, thus a brake on birth is removed. Secondly, estrogen increases the sensitivity of oxytocin receptors. **Oxytocin** is a hormone released from the posterior pituitary, which causes the smooth musle of the uterus to contract. The stimulus for oxytocin release is primarily from stretch receptors in the uterus and cervix. As the fetus begins to move down the cervix, the cervix is stretched and more oxytocin is released. This results in further uterine contractions, thus pushing the fetus further down the cervix, which is stretched further, and so on. The release of oxytocin from the posterior pituitary gland is one of the rare examples of positive feedback in biological systems. One further effect of oxytocin is the increased synthesis of a group of chemicals called **prostaglandins**. Prostaglandins are C20 lipid molecules. They have many functions, both within the female reproductive system and elsewhere. During birth they increase the contractile activity of the uterine smooth muscle, thus aiding expulsion of the fetus. Following the entry of the newborn animal into the outside world, further uterine contractions occur and the placenta is shed and expelled.

11.8 Lactation

One of the characteristic features of mammals is that the female produces milk from the mammary glands for a certain period after birth in order to provide nourishment for the newborn animal. During pregnancy, under the influence of the hormone **prolactin**, the mammary glands develop in preparation for the suckling of the newborn animal.

Suckling causes mechanical stimulation of receptors around the nipple area. Nerve impulses are conveyed back to the hypothalamic region of the brain, resulting in an increase in the release of both prolactin and oxytocin from the pituitary gland. Prolactin increases the synthesis of milk, whilst oxytocin causes the contraction of the smooth muscle surrounding the milk-producing glands, which results in ejection of milk from the mammary gland. This process is summarized in *Figure 11.14*.

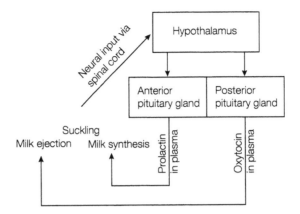

Figure 11.14. The control of lactation in the female mammal. Other stimuli, such as the sound of the offspring crying, originating in higher brain centers, may also result in milk ejection.

11.9 Maternal behavior

It is clear that the degree of parental care received by offspring from their parents will vary tremendously. For example, an insect may produce millions of fertilized eggs, and even if only a small proportion of these eggs fully develop, it would be impossible for the parent to provide a great deal of maternal care to many thousands of offspring. In contrast, primates may produce only two or three offspring who are afforded a considerable amount of maternal (and paternal) care until they are able to care for themselves. Thus, maternal care and behavior is primarily the province of higher animals.

Much remains to be discovered about the physiological mechanisms underlying maternal, and many other types of behavior. What is known is that cues from the newborn animal are important signals in initiating such behavior. It should be remembered that maternal behavior may well have been apparent before the arrival of the offspring – indeed, even before the egg was fertilized. Thus, for example, there may have been physiological and anatomical changes in order to gain a mate. The changes in the plumage of birds would be a good example of this. There may have been a complicated mating ritual and the development of nest sites, for example.

The degree of maternal behavior displayed by mothers will vary tremendously amongst even the higher animals. Thus, newborn mice and rats, which are born naked and blind, are continually kept in their nest where they are able to suckle their mother. Contrast this with the monkeys and apes, whose young are taken with the mother every time she leaves the nest. As the newly born animal grows and develops, there are changes in both its behavior and its mother's behavior. This results in the young animal becoming an independent individual.

Further reading

Cohen, J. and Massey, B. (1984) *Animal Reproduction: Parents Making Parents.* Edward Arnold, London.

Johnson, M. and Everitt (1988) *Essential Reproduction.* Blackwell, London.

Appendix. An outline classification of animals

Animals are classified into various taxonomic categories developed by the Swedish biologist Linnaeus in the nineteenth century. The categories are as follows:

Taxonomic category	Classification of humans
Kingdom	Animalia
Phylum	Chordata
Subphylum	Vertebrata
Class	Mammalia
Subclass	Euthria
Order	Primate
Family	Hominidae
Genus	*Homo*
Species	*Homo sapiens*

There are many ways to classify animals: on the basis of body symmetry, the presence of body cavities, developmental patterns, etc. The arguments for and against various classification techniques are beyond the scope of this book. What is provided below is an outline plan of the currently accepted view of animal classification. It is not intended to be an extensive account. Although treated in this book as animals on the basis of their animal-like traits, the protozoa belong to their own kingdom, the Protista, which includes other single-cell organisms; consequently, their classification is not discussed here.

1. Phylum Mesozoa
2. Phylum Placozoa
3. Phylum Porifera
    The sponges
4. Phylum Cnidaria (or Coelenterata)
    Class Hydrozoa, e.g. hydra, obelia
    Class Scyphozoa e.g. jellyfishes

        Class Anthozoa e.g. sea anemones, corals
        Class Cubozoa

5. Phylum Ctenophora
        Class Tentacular, e.g. comb jellies

6. Phylum Platyhelminthes
        Class Turbellaria, e.g. free-living planarian flatworms
        Class Trematoda, e.g. *Schistosoma* and other flukes
        Class Cestoda, e.g. *Taeniarhynchus* and other tapeworms
        Class Monogenea, e.g. monogenetic flukes

7. Phylum Nermeta
8. Phylum Gnathostomulida
9. Phylum Rotifera
10. Phylum Gastrotricha
11. Phylum Kiorhyncha
12. Phylum Nematoda
        e.g. roundworms and hookworms
13. Phylum Acanthocephala
14. Phylum Entoprocta
15. Phylum Mollusca
        Class Monoplacophora
        Class Polyplacophora
        Class Gastropoda, e.g. snails and slugs
        Class Bivalvia, e.g. clams and mussels
        Class Cephalopoda, e.g. squids and octopods
16. Phylum Annelida
        Class Polychaeta, e.g. ragworms, fanworms
        Class Oligochaeta, e.g. earthworms
        Class Hirudinea, e.g. leeches
17. Phylum Arthropoda
        Subphylum Trilobitomorpha
        Subphylum Crustacea (six major classes, including crabs and lobsters)
        Subphylum Uniramia
           Class Diplopoda, e.g. millipedes
           Class Chilopoda, e.g. centipedes
           Class Insecta, e.g. Insects
        Subphylum Chelicerata
           Class Merostomata, e.g. horseshoe crabs
           Class Arachnida, e.g. spiders and scorpions
           Class Pycnogonida, e.g. sea spiders
18. Phylum Sipuncula
19. Phylum Echiura
20. Phylum Pogonophora
21. Phylum Pentastomida
22. Phylum Onychophora
23. Phylum Tardigrada
24. Phylum Phoronida
25. Phylum Bryozoa

26. Phylum Brachiopoda
27. Phylum Echinodermata
     e.g. starfish, sea urchins and sea cucumbers
28. Phylum Chaetognatha
29. Phylum Hemichordata
30. Phylum Chordata
     Subphylum Urochordata, e.g. sea squirts
     Subphylum Cephalochordata, e.g. amphioxus
     Subphylum Vertebrata
       Class Chondrichthyes, e.g. elasmobranchs (sharks)
       Class Osteichthyes e.g. teleosts (bony fish)
       Class Amphibia, e.g. frogs and toads
       Class Reptilia, e.g. snakes and lizards
       Class Aves, e.g. birds
       Class Mammalia

Glossary

Absorption is the uptake of substances into animal cells from either the gastrointestinal tract or, in the case of some animals, from the external environment.

Actin is a myofilament protein found in muscle cells, which interacts with another protein (myosin) to cause contraction of muscle.

Action potential is a transient change in the resting membrane potential of neurons, such that the interior of the neuron becomes positively charged.

Adaptation is the phenomenon seen in sensory receptors whereby continual stimulation of a receptor results in the receptor becoming less sensitive to subsequent stimulation.

Adenylate cyclase is the membrane-bound enzyme that catalyzes the production of cAMP from ATP.

Allantois is an extra-embryonic membrane which has both a gas exchange and waste disposal function.

Alveoli are microscopic terminations of the mammalian lung and the site of gas exchange between blood and air.

Ammonotelic refers to animals who secrete the waste products of nitrogen metabolism as ammonia.

Amnion is an extra-embryonic membrane that forms a fluid-filled sac which surrounds the developing embryo.

Anticoagulants are a group of compounds which prevent the process of blood clotting.

Antidiuretic hormone (ADH) is a peptide hormone secreted from the posterior pituitary gland, which functions to control water reabsorption in the kidney.

Asexual reproduction is reproduction which occurs in a single animal without the formation of gametes. It is the result of mitotic cell division and results in the production of genetically identical offspring.

Axon is a single extension arising from the cell body of a neuron which conducts action potentials towards the axon terminals and synapses.

Axon hillock is the initial region of the axon nearest the cell body. This region has the lowest threshold value for action potential production and is the site where action potentials originate.

Axonal transport is the transport of substances between the cell body and axon terminals.

Bile is a fluid formed in the liver and released from the gall bladder, which has a detergent-like action and is important in the digestion of lipids by preventing fat droplets in the gastrointestinal tract from clumping together.

Binary fission is the method of asexual reproduction whereby the parent animal splits into two or more identical offspring.

Blastocyst is the blastula stage of mammalian embryos.

Blastomere is a cell produced as a result of the early cleavage division of the zygote.

Blastophore is the invagination of the embryo which occurs during the process of gastrulation, and which will eventually form the anus.

Blood plasma is the liquid component of blood in which the cellular elements are suspended.

Capillaries are the smallest blood vessels. They have a wall which is one cell thick and they are the site where nutrients, gases and other substances are exchanged between the blood and tissues.

Carbohydrase is the enzyme which attacks the glycosidic bond that links monosaccharides together.

Cardiac muscle is the muscle which makes up a large proportion of the heart, the contractile activity of which is responsible for forcing blood around the circulatory system.

Carnivore is an animal which eats the flesh of other animals.

Cephalization is the tendency for neurons and sensory structures to accumulate at the front of animals during the evolution of nervous systems.

Channels are protein pores in the membranes of animal cells which allow substances, primarily ions, to pass from one side of the membrane to the other. Each substance has its own unique channel.

Chemoreceptor is a sensory structure which is responsible for the detection of specific chemical substances.

Chemotaxis is the directed movement of an animal cell or whole animal in response to a specific chemical stimulus.

Chitin is a complex polysaccharide which forms part of the cuticle of the arthropods.

Chorion is an extra-embryonic membrane which forms part of the placenta in mammals.

Chorionic gonadotropin is a hormone secreted by the placenta which maintains the corpus luteum during the initial stages of pregnancy and the production of estrogen and progesterone.

Chylomicrons are the resynthesized products of fat digestion (triglycerides), which are coated with phospholipids and protein and are transferred to the lacteals of the villi during the absorption of fats from the gastrointestinal tract.

Chromatophores are cells containing pigment. The dispersal of the pigment may vary, resulting in color changes of the animal.

Chromosomes are packages of genetic material found in the nucleus of the cell. Chromosomes are formed from DNA and a number of proteins, which together are known as chromatin.

Cilia are hair-like projections originating from the membrane of some animal cells and the surface of some single-celled animals. They are motile and their role is many-fold, including the passage of substances along the cell membrane and movement of unicellular animals.

Circular muscle is muscle which is arranged around the circumference of a tubular organ.

Classical endocrine system is that part of the endocrine system which comprises endocrine organs formed from non-neural tissue, and whose secretions enter the bloodstream to influence a target organ some distance from their release.

Cleavage is the cell division which occurs to produce a multicellular zygote and embryo from a single fertilized egg.

Closed circulation is a circulatory system where blood is always contained within the blood vessels.

Coelomic cavity is a body cavity formed in animals which is lined with mesodermal tissue.

Colloid osmotic pressure is the component of the osmotic pressure of blood plasma which is caused by the presence of proteins dissolved in the plasma.

Conformational change is the change in shape of a protein molecule as a result of changes occurring at one or a few particular amino acid residues (e.g. changes in the ionization of 'R' groups, phosphorylation of an amino acid etc.).

Conduction is the transfer of heat between objects which are in physical contact with each other.

Connexon is a protein pore found at electrical synapses.

Contractile vacuole is the excretory organ found primarily in single cell animals.

Convection is the transfer of heat to a fluid (liquid or air).

Corpus luteum is the remainder of the ovarian follicle after ovulation has occurred. It secretes estrogen and progesterone.

Critical maximum temperature is the maximum external temperature an animal can tolerate before significant physiological responses are required to reduce body temperature in order to avoid death.

Critical minimum temperature is the minimum external temperature an animal can tolerate before significant physiological responses are required to increase body temperature in order to avoid death.

Crustaceans are members of the phylum Arthropoda, including animals such as crabs, lobsters, shrimps and woodlice.

Cuticle is the waxy covering found on the surface of insects.

Cyclic AMP (cAMP) is a second messenger molecule formed from ATP, the function of which is the activation of protein kinases within a cell.

Cyclic respiration is the ventilatory phenomenon seen in some insects, whereby there is continual uptake of oxygen and carbon dioxide is removed in periodic bursts.

Cytoplasm is the contents of an animal cell excluding the nucleus.

Cytoskeleton is the arrangement of filamentous proteins within a cell which represents a skeleton or scaffolding within the cell.

Deamination is the oxidative breakdown of amino acids to produce a keto acid and ammonia.

Dendrites are projections originating from the cell body of a neuron which receive inputs from other neurons and transfer local potentials towards the cell body.

Deuterostome is a group of animals, e.g. the echinoderms and vertebrates, who display radial cleavage of the embryo and in which the mouth develops after the formation of the anus. The word deuterostome means 'second mouth'.

Digestion is the break-down of ingested food into its constituent components so that they may be absorbed and utilized by the animal.

Dioecius state is the presence of separate sexes (male and female) for a given animal species.

Discontinuous respiration, see Cyclic respiration.

Ductal gland is a gland whose secretions travel via a duct to reach their final destination, e.g. bile is released into the bile duct from where it enters the small intestine of mammals.

Ductless gland is a gland whose secretions directly enter the circulating body fluids.

Eccritic temperature is the preferred body temperature of an animal.

Ectotherm is an animal which derives its heat gain primarily by absorption from the external environment.

Effector is a general term for a structure capable of producing a biological response, e.g. muscular movement, secretion of a hormone etc.

Elasmobranchs are the sharks and rays, which have a skeleton made from cartilage rather than bone.

Electrochemical equilibrium is seen in the membranes of animal cells when the concentration and electrical gradients for a particular ion are equal and opposite, and there is no net movement of the ion across the membrane.

Endocrine, see Classical endocrine system.

Endocytosis is the process by which particles are engulfed by the plasma membrane prior to being internalized by the cell.

Endoderm is one of the three layers of cells formed during gastrulation, which eventually forms the digestive and respiratory systems.

Endoparasites are parasites that live within the bodies of other animals.

Endopeptidase is an enzyme which breaks the peptide bonds linking adjacent amino acids within the interior of a peptide or protein molecule.

Endoskeleton is a skeleton which is found within animals.

Endotherm is an animal which derives its heat gain primarily from cellular metabolism.

Ephapse is an electrical synapse.

Epitoky is the mechanism of reproduction seen in some annelid worms. There is a metamorphic transformation of some of the body segments into gonads or, in some cases, free-living organisms.

Erythrocytes are red blood cells containing the oxygen transport pigment hemoglobin.

Esophagus is the hollow muscular tube which connects the mouth to the stomach.

Esterase is an enzyme which breaks down simple lipids.

Euryhaline is an aquatic animal which is capable of surviving in seawater of a wide range of salinity.

Excretory organs are the structures involved in the excretion of waste products, which ensure that the compositions of internal body fluids are correct.

Exocytosis is the fusion of a membrane-bound vesicle with the plasma membrane and the subsequent release of the contents of the vesicle to the exterior of the cell. This is the opposite of endocytosis.

Exopeptidase is an enzyme which removes the terminal amino acids from a protein molecule. There are two types: N-terminal exopeptidase and C-terminal exopeptidase.

Exoskeleton is a skeleton which is found on the outside of an animal.

Exteroreceptors are sensory receptors which monitor conditions in the external environment.

Extracellular fluid is fluid which exists outside the cell, e.g. blood plasma.

Flame cell is the excretory organ found in flatworms. It is a ciliated structure, and movement of the cilia filters the body fluids.

Fertilization is the union of male and female gametes.

Follicle-stimulating hormone (FSH) is a gonadotropic hormone released from the anterior pituitary gland.

G proteins are a group of membrane proteins that alternately bind GTP and GDP and regulate the activity of adenylate cyclase.

Gametes are sex cells, eggs or sperm, which are characterized by having only half the chromosome number of somatic (non sex cell, e.g. muscle cell) chromosomes.

Ganglia are a collection of cell bodies and dendrites in the nervous system of an animal.

Gas exchange is the transfer of gases, oxygen and carbon dioxide, between an animal and its environment.

Gastric ceca are blind-ending sacs which are connected to the lumen of a larger organ or structure.

Gastrointestinal system is the body system responsible for the ingestion, digestion and absorption of food.

Gastrula is the embryonic stage formed after the process of gastrulation.

Gastrulation is the embryonic process which transforms a blastula into a gastrula. In the case of higher animals, this results in the formation of the three layers of cells.

Generator potential is the local potential produced in specialized sensory cells as a result of stimulation.

Gene is a sequence of DNA which encodes for a particular protein.

Glial cells are a group of cells found in the nervous system of animals and whose role is to support the activities of neurons.

Glycosidic bond is the chemical bond which joins monosaccharides together to form di-, tri-, oligo- and polysaccharides.

Gonadotropin-releasing hormone (GnRH) is a hormone released from the hypothalamus that controls the subsequent release of the gonadotropins from the anterior pituitary gland.

Gonads are organs that produce the gametes, i.e. the testes and ovaries.

Guanosine diphosphate (GDP) is the guanosine analog of ADP which is bound to G proteins when they are at rest.

Guanosine triphosphate (GTP) is the guanosine analog of ATP which becomes bound to G proteins in exchange for GDP when the G proteins are activated.

Gular fluttering is a form of panting seen in some birds, which is a means of losing heat and therefore maintaining an appropriate body temperature.

Hematocrit is the proportion of blood occupied by red blood cells.

Hemocoel is a cavity, similar to the coelom in higher animals, which contains hemolymph.

Hemolymph is the fluid in the body cavities of invertebrates with an open circulation. It is equivalent to blood seen in higher animals.

Herbivorous describes an animal whose diet comprises plants.

Hermaphrodite is an animal which is simultaneously both male and female.

Heterotrophs are organisms that are dependent upon their environment for the supply of preformed nutrients. This definition includes all animals. This contrasts with autotrophs, i.e. green plants, which use the process of photosynthesis to make organic molecules (glucose) from inorganic molecules (carbon dioxide and water).

Hibernation is the phenomenon seen in some mammals, which enter a period of inactivity during the winter months that is associated with a massive drop in metabolic rate and body temperature.

Histamine is an amino acid derivative which functions as a local hormone in mammals. Its release from stores is increased during allergic reactions.

Holoblastic is a type of cleavage pattern observed in some fertilized eggs. The cleavage lines which are necessary for cell division pass through the entire cell.

Holometabolous development is the type of development seen in some insects, for example, where there is complete metamorphosis of the developing animal. A good example would be the change from caterpillar to butterfly.

Homeostasis is the maintenance of a relatively constant internal environment in animals.

Hormones are a chemically diverse group of substances, released from a variety of glands, which represent a means of controlling and coordinating the physiology of an animal.

Hyperosmotic is a solution which has a greater osmotic potential (i.e. is more concentrated) than another with which it is being compared.

Hypertonic is a solution which has a greater tonicity than another with which it is being compared.

Hypoosmotic is a solution which has a lower osmotic potential (i.e. is more dilute) than another with which it is being compared.

Hypothalamic pertains to the hypothalamus, a region of the mammalian brain where there is close interaction between the nervous and endocrine systems.

Hypotonic is a solution which has a lower tonicity than another with which it is being compared.

Insulin is a pancreatic protein hormone which reduces plasma glucose levels when they are elevated, e.g. immediately after a meal.

Intercalated disks are plasma membrane regions between adjacent cardiac muscle cells. They represent regions of low electrical resistance, and allow action potentials to pass from one cell to the next, ensuring that the muscle cells contract synchronously.

Intermediate filaments are a group of proteins which stabilize cell structure and resist movement.

Interoreceptors are sensory receptors which monitor conditions inside the animal.

Intracellular fluid is the fluid found on the interior of cells.

Isoosmotic is a solution which has the same osmotic pressure (i.e. the same concentration) as a solution with which it is being compared.

Isotonic is a solution which has the same tonicity (i.e. the same composition) as a solution with which it is being compared.

Lacteal is a blind-ending sac in villi into which the products of fat absorption are transported prior to entering the lymphatic circulation and the systemic venous circulation.

Lateral line is a line of sensory mechanoreceptors along the side of the body of fishes, and also amphibians, which are used for detecting movement.

Leukocytes are white blood cells with a variety of defence roles.

Lipase is an enzyme which breaks down lipids such as triglycerides.

Local current is the movement of positive charge towards negative charge, such as occurs on the interior of axons, resulting in the depolarization of adjacent regions of the axon and the passage of the action potential down the length of the axon.

Local potentials are small changes in membrane potential, either depolarizing or hyperpolarizing, that are short-lived and are not propagated from their point of origin.

Longitudinal muscle is muscle that is arranged along the length of a tubular organ.

Loop of Henle is a region of the nephron of a kidney between the proximal and distal convoluted tubules where the production of concentration gradients allows the kidney to produce a concentrated urine.

Malpighian body is the glomerular capillaries and Bowmans capsule found at the beginning of the nephron.

Malpighian tubules are blind-ending sacs which are the excretory organs of insects.

Meiotic cell division is cell division which results in the production of four daughter cells, each of which has half the chromosome number of the original parent cell.

Meroblastic is a cleavage pattern seen in some fertilized eggs. The cleavage lines required for cell division are unable to pass through the large amount of yolk seen in some eggs.

Mesencephalon is one of the brain vesicles seen in the developing vertebrate embryo which will ultimately form part of the midbrain in the adult vertebrate.

Mesoderm is one of the layers of body cells formed during gastrulation which will eventually form the muscular, cardiovascular and other organ systems in the fully developed animal.

Metabolic rate is a measurement of the sum total of all metabolism occurring in an animal at any given time.

Metabolic water is water formed as a result of the oxidation of fats and lipids in the animal as part of its metabolism.

Metamorphosis is the change in appearance of some free-living animals which occurs during their development into the adult form, e.g. the change from a caterpillar to butterfly which completes the development of this particular animal.

Metanephridia are the excretory organs found, for example, in the annelid worms.

Microfilaments are a group of proteins, including actin and myosin, which are responsible for cell shortening, e.g. during muscle contraction.

Microtubules are polymers of a protein called tubulin which are required by cilia and flagella.

Mitotic cell division is cell division which results in the formation of genetically identical daughter cells, all of which have the same chromosome number as the parent cell.

Monoecious describes animals which possess both male and female gonads, essentially the same as hermaphroditism.

Morula is a solid mass of cells formed by division of the fertilized egg early in the development of the zygote.

Mosaic development is a type of development where the final fate of each cell is predetermined from early on in the developmental process.

Motor unit is a motor neuron plus all the muscle fibers which it innervates.

Myelin sheath is a fatty sheath formed by glial cells, which wraps around the axons of some neurons. Its role is to increase the velocity at which action potentials travel along the axon.

Myofibrils are the contractile elements which form muscle fibres.

Myofilaments are the proteins, e.g. actin and myosin, which form the myofibrils.

Myosin is one of the contractile proteins found in muscle cells.

Negative feedback is observed when the initial disturbance in a physiological system results in inhibition of the input that caused the disturbance, thus returning the system to normal.

Nephridiopore is the exit to the environment of the excretory organ, the metanephridia, found in annelid worms.

Nephridiostome is the ciliated opening to the metanephridia in annelid worms.

Neuroendocrine system is the type of endocrine system in which neurotransmitters are released into circulating body fluids and may have a site of action some distance from their release. This contrasts with the local action that neurotransmitters usually have in the synapse.

Neurohormones are the neurotransmitters released in the neuroendocrine system.

Neurons are the functional cells of the nervous system. They work by generating, transmitting and receiving action potentials.

Neuropeptide is a peptide that is released from nerve endings and may function either as a neurotransmitter or as a neuromodulator.

Neurotransmitter is a chemical substance released from axon terminals which interacts with and alters the electrical activity of other cells, e.g. neurons and muscle cells.

Odontophore is a tooth-like structure in molluscs on which the radula sits.

Olfaction is the sense of smell.

Oligosaccharide is a carbohydrate formed from the joining together of a small number of individual monosaccharides.

Open circulation is an arrangement where blood is pumped out of the heart into blood vessels. It then leaves the vessels of the circulatory system and is distributed through the tissue spaces before returning to the heart to be pumped out again.

Osmoconformers are animals which maintain the osmotic potential of their blood at the same level as that of the environment in which they live.

Osmoregulation is the ability to regulate the composition of body fluids in terms of both solute and solvent concentration.

Osmoregulators are animals which maintain the osmotic potential of their blood at a different level to that of the environment in which they live.

Osmosis is the movement of water across a selectively permeable membrane, from a region of high water concentration (i.e. a dilute solution) to a region of low water concentration (i.e. a concentrated solution).

Osmotic equilibrium is the situation seen when two solutions have the same osmotic pressure and there is no net movement of water from one solution to the other.

Osmotic pressure is the pressure which must be applied to a solution in order to prevent water entering it by osmosis.

Ovary is the female gonad which produces eggs.

Oviparous describes a reproductive strategy in which fertilized eggs develop outside the mother's body.

Ovoviviparous describes a reproductive strategy in which eggs are fertilized and develop internally. The embryo obtains all its nutrients from the egg, which then hatches either internally or immediately after it has been laid.

Pacemaker is a cell or group of cells which sets a rate to which all other cells adhere, e.g. heartbeat frequency and the frequency at which cilia beat are determined by pacemakers.

Parabronchi are the functional units of the bird lung and the site where gas exchange occurs.

Paracrine is a local hormone which does not enter the circulating body fluids to reach its target organ.

Parthenogenesis is the development of an unfertilized egg. All the resultant offspring will be female, as were the eggs from which they developed (i.e. they will all contain the female sex chromosomes).

Partial pressure is the pressure any given gas in a mixture of gases contributes to the total pressure of the gas mixture.

Peptide bonds are the chemical bonds which link individual amino acids in a peptide or protein molecule.

Peristalsis is the movement of matter along the gastrointestinal tract by the alternate contraction and relaxation of the circular and longi- tudinal muscle which surrounds it.

Pheromones are chemical substances (hormones) which are released into the environment in order to communicate with a member of the same species.

Pituitary gland is a small endocrine gland at the base of the vertebrate brain.

Placenta is a structure formed between tissues of the mother and tissues of the embryo which serves to supply the developing embryo with nutrients.

Plasma, see Blood plasma.

Plasmagel is the viscous type of cytoplasm around the periphery of the cell.

Plasmasol is the liquid type of cytoplasm found in the cell.

Positive feedback is the situation observed whereby the initial distur- bance in a physiological system results in further disturbance.

Postsynaptic potentials, see Local potentials.

Prosencephalon is the forebrain of vertebrates.

Protease is the general term for an enzyme which breaks down proteins.

Protonephridia are the excretory organs found in the platyhelminthes.

Protostomes are a group of animals, including the annelids and the arthropods, that undergo spiral cleavage of the zygote, during which the mouth develops first.

Q_{10} is the ratio between the rate of a biological process at any given temperature and at 10°C above that temperature, i.e. Q_{10} = rate at (T+10)°C/rate at T°C.

Radial cleavage is a cleavage pattern seen in some developing zygotes which results in layers of cells (blastomeres) lying directly on top of each other.

Radiation is heat transfer between objects which are not in direct phys- ical contact with each other. It relies on the fact that all objects emit electromagnetic radiation of a particular wavelength and intensity. These two factors in turn depend upon the temperature of the object concerned.

Radula is a file-like tongue found in molluscs which is used for scraping small particles of food off a larger particle.

Receptor potential is the local potential produced in a sensory nerve ending as a result of appropriate stimulation, which may, if the stimulus is great enough, generate an action potential. This is analogous to the generator potentials produced in specialized sensory cells.

Refractory period is the period of time immediately after one action potential has been generated, when a neuron is unable to generate a second action potential. The refractory period may be split into two parts — an absolute and a relative refractory period.

Regulative development is a type of development where the final fate of each cell is not predetermined from early on in the developmental process.

Relative humidity is a measure of the amount of water in air. When air is fully saturated with water, the relative humidity is said to be 100%; when air contains no water vapor, the relative humidity is said to be 0%.

Relaxin is a peptide hormone whose role is to soften the tissues of the cervix prior to birth in mammals.

Resting membrane potential is the potential difference (voltage) measured across the plasma membrane of all animal cells, but which is largest in neurons. In this case, there is a voltage of 70 mV across the membrane, the inside being negative with respect to the outside.

Rhodopsin is a photopigment which captures photons of light and transduces them into electrical activity in the rods of the vertebrate eye.

Rhomencephalon is an early embryonic brain vesicle which forms the brainstem and cerebellum in adult vertebrates.

Saliva is an aqueous solution containing an enzyme (amylase) and mucins secreted by the salivary glands into the mouth.

Salt glands are glands seen in a variety of animals, e.g. fish and birds, which secrete solutions rich in sodium chloride. They contribute to the process of osmoregulation in these animals.

Sarcolemma is the plasma membrane which surrounds muscle cells.

Sarcoplasm is the cytoplasm of muscle cells.

Sarcoplasmic reticulum is the endoplasmic reticulum of muscle cells.

Second messengers are molecules released into a target cell in response to the stimulus of the first messenger, e.g. a hormone, and whose role is to initiate further intracellular responses.

Sensory receptors are a group of structures, either specialized cells or simple nerve endings, which monitor changes in both the internal and external environment of animals.

Sertoli cells are cells in the testes associated with the process of spermatogenesis.

Sexual reproduction is reproduction which requires male and female parents to produce sperm and eggs, respectively. The union of egg and sperm results in the formation of a fertilized egg which then undergoes further development to produce viable offspring.

Skeletal muscle is muscle which is striated (striped) in appearance and which is under voluntary control in vertebrates. Contraction of this muscle results in muscle movement of the animal.

Smooth muscle is muscle which has no visible striations. In mammals this muscle is not under voluntary control. Smooth muscle surrounds organs such as the gastrointestinal tract, and it is the contraction of this muscle which is responsible for peristalsis in this system.

Solenocyte is a cell containing a single cilia that forms the end of the protonephridia in platyhelminthes.

Spermatogonia are undifferentiated cells that will eventually develop into mature sperm cells.

Spiracles are small openings in the body wall of insects which link the tracheal system to the environment.

Spiral cleavage is a type of cleavage in which groups of cells are produced that lie in the grooves between the cells below them rather than directly on top of them.

Stenohaline describes an aquatic animal that can only tolerate a narrow range of salinity.

Striated muscle, see Skeletal muscle.

Stolonization is the process of reproduction by budding seen in some animals.

Summation is the ability of a number of postsynaptic potentials to produce a cumulative effect which may or may not generate an action potential.

Suspension feeding is the removal of foodstuffs from large volumes of fluid (air or water) by simple filtration of the fluid.

Symbiotic describes a relationship where two or more organisms live together for the mutual benefit of all.

Synapse is the gap between an axon terminal and a second cell, e.g. another neuron or muscle cell.

Teleost is a fish which has a skeleton made of bone.

Thermal tolerance range is the range over which body temperature is at an acceptable level.

Thermoreceptors are sensory receptors which monitor the temperature of the internal and external environment of an animal.

Threshold is the value to which resting membrane potential must be raised before an action potential may be generated.

Thrombocyte is an alternative term for a group of blood cells known as platelets. Their role is to form temporary plugs in holes which appear in blood vessels, thus preventing excessive blood loss.

Thyroxine is a hormone produced by the thyroid gland. It is a derivative of the amino acid tyrosine and has several functions in different animals.

Tonicity refers to the response seen in an animal cell when placed in a particular solution. A cell will lose water and shrink when placed in a hypertonic solution and will gain water and burst when placed in a hypotonic solution. This is because of the differing compositions of the intracellular fluid and the solution in which the cell is placed.

Trachea is a large, hollow tube through which air moves as it passes deeper into gas exchange organs.

Tracheoles are small branches of the trachea found in insects, which carry air into the insect's body. Occasionally, tracheoles may deliver air to individual cells.

Trimethylamine oxide (TMAO) is a nitrogen-containing substance which is added to the plasma of elasmobranchs in order to increase the osmotic concentration of blood so that it is equal to that of seawater.

Trophoblast is the outer layer of cells of the mammalian blastocyst. The cells secrete a proteolytic enzyme which allows the blastocyst to embed in the endometrium.

Tropomyosin is a so-called regulatory protein found in muscle cells. It blocks the binding sites for myosin on the actin filaments.

Troponin is a regulatory protein found in muscle cells. This is the protein to which calcium binds in order to initiate muscle contraction.

Ultrafiltration is the filtration of body fluids which results in the formation of a protein-free filtrate, i.e. the filter allows all molecules, with the exception of proteins, to pass through it.

Unmyelinated neurons are neurons which lack a myelin sheath around the axon.

Urea is a nitrogenous metabolic waste product which is formed from ammonia. It is reasonably soluble so it can be excreted in urine, but is far less toxic than ammonia.

Ureotelic describes animals which excrete their nitrogenous waste in the form of urea.

Uric acid is a nitrogenous metabolic waste product which is formed from ammonia. It is extremely insoluble in water and is excreted by those animals in which water loss must be minimized.

Uricotelic describes animals which excrete their nitrogenous waste in the form of uric acid.

Urochordates are a primitive group of chordates. They possess a notochord (a primitive type of endoskeleton) which is a characteristic feature of this group of animals. They include such animals as the sea squirts.

Ventilation:perfusion ratio is an indicator of the efficiency of gas exchange. The ventilation:perfusion ratio is the ratio between the volume of oxygen entering a gas exchange organ and the volume of blood passing through it.

Vesicular eye is the type of eye seen in some molluscs and vertebrates, for example. It possesses a single lens, the movement of which ensures that objects are correctly focused on the retina.

Vestibular system is the sensory system which provides animals with information regarding their position in space, e.g. which way up they are and whether they are moving or stationary.

Viviparity is the reproductive strategy whereby fertilization and development are internal. Nutrients required for the development of the embryo are provided by the mother and the offspring are born relatively well developed.

Yolk sac is one of the extra-embryonic membranes of vertebrates.

It contains yolk, which can be a vital source of nutrients to the developing zygote.

Zygote is the fertilized egg produced as a consequence of the union of egg and sperm.

Zymogen is the general term for the inactive precursor of a proteolytic enzyme.

Index

Abomasum, 136
Absorption amino acids, 144
Absorption carbohydrates, 143
Absorption lipids, 145
Absorption mechanisms, 140–142
Absorption structural adaptations, 142–143
Acclimatization, 8
Acetylcholine, 23–24, 50, 113
Actin, 46–52
Action potential, 5, 9, 11, 13, 16–20, 26, 34, 55
Adaptation
  phasic, 36
  tonic, 36
Adenosine, 23
Adenylate cyclase, 60
Adrenal cortex, 70
Adrenal medulla 70
Adrenaline (epinephrine), 58, 70
Adrenocorticotropic hormone, 71
After-hyperpolarization, 18
Air sacs, 84–85
Aldosterone, 70
Alveoli, 84
Aminopeptidases, 139
Ammonia, 171
Ammonotelic, 171
Amoeboid movement, 46
Amylase, 134
Androgens, 70
Antennal gland, 166
Anticoagulants, 134
Antidiuretic hormone (ADH), 23, 56, 58, 68, 69, 170
Antifreeze molecules, 168
Antifreeze proteins, 98
Asexual reproduction, 175, 177
Aspartate, 23
Atrial naturetic peptide, 71
Axon, 11, 16, 18–19, 68
Axon hillock, 11, 18, 24
Axon terminals, 45
Axonal transport, 45

Bile, 140
Binary fission, 175–176
Bipolar neuron, 12
Bladder, 158
Blastocyst, 188
Blastomere, 185
Blood coagulation, 4
Blood composition, 108
Blood oxygen, 8
Blood plasma, 1
Blood vessels, 115
  arteries, 116
  arterioles, 116
  capillaries, 116
  venules, 116
Body fluid composition, 151
Body temperature, 4–8, 40, 102
Bohr effect, 122
Bohr effect magnitude, 122
Bombykol, 37
Bowmans capsule, 167
Brain, 5, 13, 28–31,
Breathing (ventilation) control, 87
  and carbon dioxide, 89
  and oxygen, 89
Brown adipose tissue, 100
Budding, 176

cAMP, 60–61
Carbamino compounds, 124
Carbohydrases, 136
Carbon dioxide dissociation curve, 124
Carbon dioxide solubility in water, 76
Carbon dioxide transport, 123
Carbonic anhydrase, 123
Carboxypeptidases, 140
Cardiac muscle, 47, 51, 111
Carnivores, 128
Cellulose digestion, 131, 135–136
Central nervous system, 29
Cephalization, 28
Cerebrospinal fluid, 88
Chemoreception, 37
Chemoreceptors, 34

Chemotaxis, 37
Chitin, 54, 157
Chloride cells, 154
Chlorocruorin, 119
Chorionic gonadotropin, 189
Chorionic villi, 188–189
Chromatophores, 44, 65
Chylomicrons, 145
Chyme, 135–136
Chymotrypsin, 139
Chymotrypsinogen, 139
Cilia, 26–27, 38–39
  and feeding, 126–127
Circulatory system
  closed, 107, 115
  double, 117–118
  open, 107, 114, 165
  single, 117
Classical endocrine system, 56
Cleavage,
  radial, 185
  spiral, 185
Coelom, 110
Collecting duct, 170
Colloid osmotic pressure, 108
Compound eye, 40
Connexon, 20–21
Contractile vacuoles, 161–162
Copora allata, 66
Corpus luteum, 189
Corticosterone, 70
Corticotropin (adrenocortico-tropic hormone, ACTH), 69–70
Corticotropin releasing hormone (CRH), 69–70
Critical maximum temperature, 98
Critical minimum temperature, 98
Cross bridge, 49
Cuticle, 157
Cytoskeleton, 45

Deamination, 171
Dendrites, 10–11
Depolarization, 5, 17, 19, 35

Deuterostomes, 186
Development,
  mosaic, 186
  regulative, 186
Diacylgycerol, 61
Dioecious state, 177
Discontinuous (cyclic)
  respiration, 87
Dopamine, 23, 113

Eccritic temperature, 98
Ecdysone, 58, 67
Echolocation, 38
Ectoderm, 10, 186
Ectotherm, 92
  aquatic, 95
  terrestrial, 97
Ectothermy and endothermy,
  105
Endoparasites, 130
Effector, 5–6, 43
Egg (ova) production,
  181–182
Electroreception, 42
Electroreceptors, 34
Endocrine system, 55–56
  invertebrate, 62–67
Endocrine organs, identification,
  57
Endoderm, 110, 186
Endometrium, 188
Endopeptidases, 138
Endoskeletons, 54
Endotherm, 92, 99
  body temperature, 100
Enzymes, 7
Ephapse, 20–21
Epitoky, 63
Erythrocytes (red blood cells),
  108–110
Esophagus, 134
Esterases, 140
Estrogens, 70
Estrus cycle, 183
Euryhaline, 150
Excretory organs,
  general, 161
  specialized, 161
Exocytosis, 21
Exopeptidases, 138–139
Exoskeleton, 53–54, 66
Exteroreceptors, 34
Extracellular digestion, 131
Extracellular fluid, 1–3, 13–15,
  147
Extraembryonic membranes,
  allantois, 188
  amnion, 188–189
  chorion, 188–189
  yolk sac, 188–189

Feces, 146, 157–158
Feedforward, 7
Feeding, 26
Feeding methods, 126
Fertilization, 177, 183
  external, 183
  internal, 183
Ficks law, 75
Flame cell, 162–163
FMRFamide, 23
Follicle-stimulating hormone
  (FSH), 69, 71, 180–181
Follicle-stimulating hormone
  releasing hormone
  (FSHRH), 69
Food vacuole, 131

G protein, 59
GABA (gamma-aminobutyric
  acid), 23, 113
Gametes, 175, 177, 183
Ganglia, 29
Gas exchange, 73
  across body surface, 76
  organs design, 77
  organs respiratory medium
  and body fluids, 78
Gases in water, 74
Gastrin, 70
Gastrointestinal system regions,
  132–133
Gastrula, 186–187
Gastrulation, 186
Generator potential, 36
Gill filaments, 81
Gill lamella, 81
Gills,
  anal, 155
  and excretion, 161
  and feeding, 126–127
  and ions, 152–153,
  155–156
  fish, 82
  heat loss, 96
  invertebrates, 79
  vertebrate, 80
Gizzard, 134
Glial cell, 9, 13, 24
Glomerulus, 167
Glucocorticoids, 70
Glutamate, 23
Glycosidic bonds, 136
Goldman equation, 16
Gonadotropin releasing
  hormone (GnRH),
  180–181
Gonads, 177
Growth hormone, 69, 71
Growth hormone-releasing
  hormone (GHRH), 69

Growth hormone release-
  inhibiting hormone
  (somatostatin, GHRIH), 69
Guanosine diphosphate, 59
Guanosine triphosphate, 59
Gular fluttering, 103, 159

Haldane effect, 124
Heart, 111
  branchial, 118
  chambered, 112
  myogenic, 113
  neurogenic, 113
  pacemaker, 113
  tubular, 112
Heat exchangers, 104
Heat transfer
  conduction, 93
  convection, 94
  evaporation, 95
  radiation, 94
Hematocrit, 109
Hemerythrin, 119
Hemimetabolous development,
  67
Hemocoel, 107, 164
Hemocyanin, 119
Hemoglobin, 119
  oxygen transport, 120
Hemolymph, 1, 107
Henry's law, 75
Herbivores, 128, 131, 135
Hermaphrodite, 178
Heterotrophs, 125
Hibernation, 102–103
Histamine, 23, 56–57
Holometabolous development,
  67
Homeostasis, 1–4, 6–8, 104
Homeothermic, 92
Hormone action, 59, 60
Hormones, chemical nature,
  58
Hydrostatic skeletons, 53
Hyperosmotic, 148
Hyperpolarization, 19, 24, 35
Hyperthermia, 104
Hypertonic, 149
Hypoosmotic, 148
Hypothalamus, 5, 67–70, 104
Hypotonic, 149

Inhibin, 180
Inositol triphosphate, 61
Insulin, 45
Intercalated discs, 47
Intermediate filaments, 45
Internal digestion, 131
Interneuron, 12, 20
Interoreceptors, 34

Intracellular fluid, 3, 13–15, 147
Ion channel, 17, 21, 24, 35, 38, 60
Isolecithal eggs, 185
Isosmotic, 149
Isotonic, 149

Jacobson's organ, 37
Juvenile hormone, 67

Kidney, 153

Lactation, 190–191
Lacteals, 145
Large intestine, 145
Lateral line, 38–39
Leukocytes (white blood cells),
    109–110
Leydig cells, 181
Lipases, 136, 140
Liver, 161
Local current, 18, 38
Local potential, 24–25, 36
Loop of Henle, 169–170
Lungs, 8, 81
    book, 83
    invertebrate, 82
    vertebrate, 84
Luteinizing hormone (LH), 69,
    71, 180
Luteinizing hormone-releasing
    hormone (LHRH), 69

Magnetoreception, 42–43
Magnetoreceptors, 34
Malpighian body, 167
Malpighian tubules, 157, 161,
    164–165
Maternal behavior, 191
Mauthner cells, 20
Mechanoreceptors, 34–35, 38
Melanocyte-stimulating
    hormone (MSH), 69, 71
Melanocyte-stimulating
    hormone release-inhibiting
    hormone (MSHRIH), 69
Melanocyte-stimulating
    hormone-releasing hormone
    (MSHRH), 69
Menstrual cycle, 183
Mesencephalon, 31
Mesoderm, 110, 186
Mesolecithal eggs, 185
Metabolic heat production, 100
    thyroid hormones, 101
Metabolic water, 158, 160
Metamorphosis, 67, 71
Metanephridia, 163–164
Microfilaments, 45
Microtubules, 45
Mineralocorticoids, 70

Molt-inhibiting hormone (MIH),
    56
Molting, 54, 66
Monoecious, 178
Morula, 185
Motor neuron, 12
Motor unit, 50
Multipolar neuron, 11
Muscle, 47
Myelin sheath, 11
Myelinated axon, 16, 18
Myofibril, 48–49
Myofilaments, 48
Myosin, 46, 48–52

Na^+/K^+-ATPase pump, 18, 101
Negative feedback, 4, 7
Nephridial canal, 166
Nephridiopore, 164
Nephridiostome, 164
Nephron, 167–169
Nernst equation, 15, 17
Nerve cells, 5
Nerve cord, 28
Nerve net, 27–28
Nervous system, 9, 30, 32, 35, 67
Neuroendocrine control, 57
Neuroendocrine system, 56, 62
Neurohormones, 56
Neuromodulator, 23
Neuron, 9, 10, 11, 13, 18, 24,
    26–27
Neuropeptides, 11
Neurosecretory cells
    lateral, 66
    median, 66
    subesophageal, 66
Neurotransmitter, 10, 21–24, 36,
    45, 50, 55
Nitrogen metabolism, 171
Node of Ranvier, 11, 18–19, 36
Nonmyelinated (unmyelinated),
    axons, 16
Noradrenaline (norepinephrine),
    23, 113–114

Odontophore, 128
Olfaction, 37
Omasum, 136
Ommatidia, 40–41
Osmoconformers, 108, 149–150
Osmoregulation, 147
Osmoregulators, 108, 149–150
Osmosis, 147–149, 166
Osmotic concentration,
    148–149
Osmotic eqilibrium, 3, 7
Osmotic pressure, 148
Ostia, 114
Ovary, 70, 181–182

Oviparity, 181
Ovoviviparous, 181
Oxygen, 73–74
Oxygen content of water, 76
Oxygen–hemoglobin
    dissociation curve, 121
Oxygen transport, 119
Oxytocin, 68–69, 190

Pacemaker,
    cilia, 27
Pacinian corpuscle, 34–35
Pain, 36
Pancreas, 137
Panting, 102–103
Parabronchi, 84
Paracrine, 56
Parathormone (calcitonin), 70
Parathyroid gland, 70
Parthenogenesis, 178
Partial pressure of gases, 74
Penis, 183
Pepsin, 138–139
Peptide bonds, 138
Pericardial organ, 65
Peripheral nervous system, 29
Peristalsis, 134
pH, 2–3, 5–7
Phagocytosis, 126
Pheromones, 37
Photoreceptor, 34–35, 40, 42
Physiological color change, 97
Pituitary gland, 56, 67–70, 170,
    190
Placenta, 188–189
Plasma, 108–109
Plasmagel, 46
Plasmasol, 46
Pneumostome, 83
Poikilothermic, 2
Positive feedback, 4–5, 17
Postcommisural organ, 64–65
Postsynaptic neuron, 20–21
Postsynaptic potentials, 24
Potassium channel, 14
Presynaptic neuron, 20–21
Proctolin, 23
Progestogens, 70
Progesterone/estrogen ratio in
    birth, 190
Prolactin, 69, 190
Prolactin release-inhibiting
    hormone (PRIH), 69
Prolactin-releasing hormone
    (PRH), 69
Prosencephalon, 31
Prostaglandin E2 58
Prostaglandins, 190
Protease, 135–136, 138
Prothoracic gland, 66

Prothoracicotropic hormone
(PTTH), 67
Protonephridia, 162
Protostomes, 186
Pseudopodium, 46

Q10, 91

Radula, 128–129
Ram ventilation, 81
Receptor, 5, 21, 59
Receptor potentials, 36
Rectal gland, 153
Refractory period,
absolute, 19
relative, 19
Regional heterothermy, 102
Release-inhibiting hormones, 69
Releasing hormones, 69
Renal tubule, 167–168
Respiratory pigments, 119
Resting membrane potential,
13–17, 113
Reticulum, 135
Retina, 42–43
Rhodopsin, 40
Rhombencephalon, 31
Relaxin, 189
Root effect, 122
Rumen, 135
Ruminants, 135

Saliva, 134
Salt glands, 159, 161
Saltatary conduction, 18
Sarcolemma, 48
Sarcomere, 48–49
Sarcoplasm, 48
Sarcoplasmic reticulum, 48, 50,
52
Seawater composition, 151
Second messenger, 22, 61
Seminiferous tubules, 179
Sensory neuron, 12
Sensory receptors, 10, 32, 34–35
Serotonin (5-HT), 23, 113
Sertoli cells, 180–181
Sexual reproduction, 175
Shivering, 100

Short cardiac peptide, 23
Sinus gland complex, 64–65
Skeletal muscle, 48–49
Skeletal muscle, fast fibres, 51
Skeletal muscle, slow fibres, 51
Smooth muscle, 48
Smooth muscle, multi unit, 47
Smooth muscle, single unit, 47
Sodium channel, 14
Solenocyte, 162
Sperm, 178, 180
Sperm production, 180
Spermatogonia, 179
Spinal cord, 5, 30
Spiracles, 86–87, 157
Stenohaline, 150
Steroid hormones action,
61–62
Stolonization, 64, 176
Stomach, 70, 134–135
Striated muscle, 47
Substance P, 23
Summation,
temporal, 25
spatial, 25
Super cooling, 98
Suspension feeding, 126
Sweat, 94, 99, 102
Sweating, 103
Symbiotic partnerships and
food, 130
Synapse, 11, 20–24, 56
Synapsin, 21–22

Telolecithal eggs, 185
Testes, 70, 178, 180
Testosterone, 70, 180
Tetany, 51
Thermal tolerance range, 98
Thermoconformers, 92
Thermoreceptor, 6, 34, 39–40,
104
central, 104
cold, 104
hot, 104
peripheral, 104
Thermoregulation, 6
Thermoregulators, 92
Threshold, 17–19, 24

Thrombocytes, 109
Thyroid gland, 71
and heat production, 101
Thyrotropin-releasing hormone
(TRH), 69
Thyrotropin-stimulating
hormone (TSH), 69
Tonicity, 149
Tooth function, 129
Trachea, 85
Tracheal system, 78, 80, 85–86
Tracheoles, 85
Trimethylamine oxide (TMAO),
152–153
Trophoblast, 188
Tropomyosin, 48 , 50
Troponin, 48, 50
Trypsinogen, 139
Trypsin, 139

Ultrafiltration, 163
Umbilical cord, 188
Unipolar neuron, 12
Unmyelinated (nonmyelinated),
neurons, 32
Urea, 152–153, 172
Urea cycle, 172
Ureotelic, 172
Uric acid, 158, 165–166, 173
Uricotelic, 173
Urine, 153, 155, 157, 163, 165,
168, 169, 170

Vasoconstriction, 101–102
Vasodilation, 102
Ventilation–perfusion ratio,
78
Vesicular eye, 40
Vestibular system, 32
Viviparity, 181

Water loss and gain, 157

Y organ, 65

Zoochloella, 131
Zooxanthellae, 131
Zygote, 177, 185
Zymogen, 138